SCIENCE SHADOWLANDS

LIGHT FOR THE PATHWAY HOME

NEIL SAMPSON

Copyright 2024
Neil Sampson
gardensgate2000@gmail.com

Cover Design: Linda Lanning

ISBN : 978-1-939456-38-0
Library of Congress number pending

Printed in USA

First Printing - October 2024

E-mail - truth@searchforthetruth.net
Web - www.searchforthetruth.net
Mail - 3255 Monroe Rd.; Midland, MI 48642

Dedication

For all who've been told it's too dark.

Publisher's Note

Life is full of contradictions.

It was a theological student (Charles Darwin) who destroyed the faith of hundreds of millions. And a horticulturist without a Ph.D., from a tiny Canadian province, who has pulled back the curtain on the Wizard of Oz assumptions of modern science. Sometimes those closest to a problem are the least likely to find a solution. Backing away from the trees allows us to see the forest.

I receive and read dozens of books on the evidence for creation each year, but this small volume is the most concise, thought-provoking, and unique treatise on the topic I have ever encountered. Neil Sampson is a wordsmith who understands that words are like currency – not meant to be wasted lest they lose their value.

This book was originally published as Lanterns on the Nashwaak: Reflections to Light the Way Home. The Nashwaak (Nash'walk) is a tributary of the St. John River which flows through western New Brunswick.

Individual reflections from this book won both first AND second place in New Brunswick province's writing competition, in BOTH 2020 and 2021. You'll discover why as you read even one of these short observations on life and scientific reality.

The author has permitted Search for the Truth Publications to distribute his book in the hopes that others … still lost on the trail … will find the blazed trees that mark the way home.

Bruce Malone, Search for the Truth Publications

"… to seek out … the reason of things …"

Eccl. 7:25

Contents

Preface i

Introduction iii

Reflection 1 – Mind 1

Reflection 2 – Wonder 4

Reflection 3 – Conscience 7

Reflection 4 – Origin 9

Reflection 5 – Life 12

Reflection 6 – Simple Cells 15

Reflection 7 – DNA 18

Reflection 8 – Mutations 21

Reflection 9 – Speciation 25

Reflection 10 – Second Law of Thermodynamics . . . 29

Reflection 11 – Genetic Entropy 32

Reflection 12 – Evolution 35

Reflection 13 – Natural Selection 39

Reflection 14 – Variation 42

Reflection 15 – Vestigial Structures 46

Reflection 16 – Irreducible Complexity 48

Reflection 17 – Eyes and Ears 51

Reflection 18 – Homology 53

Reflection 19 – Fossils 55

Reflection 20 – Boneyards 57

Reflection 21 – Catastrophism 59

Reflection 22 – Uniformitarianism 61

Reflection 23 – Cambrian Explosion 65

Reflection 24 – Transitional Forms 69

Reflection 25 – Geology 72

Reflection 26 – Carbon Dating 75

Reflection 27 – Scientific Method 78

Reflection 28 – Unfalsifiability 82

Reflection 29 – Design 86

Reflection 30 – Authority 89

Reflection 31 – Conclusion 1 92

Reflection 32 – Conclusion 2 95

Reflection 33 – Conclusion 3 98

Reflection 34 – So … where to from here? 101

Reflection 35 – A most unsavoury blend 104

Reflection 36 – A fork in the road 111

Reflection 37 – I'll have a large grapenut, please . . . 113

Reflection 38 – Look at that! I close my eyes
                and it disappears 118

Reflection 39 – I didn't know that was in there 120

Epilogue 123

Acknowledgments 124

About the Author 125

Resources 126

Preface

I am not a trained zoologist. I have no degrees in geology, anthropology, or in a dozen other scientific disciplines. As to the purchasing of online diplomas, I admit I have not botany.

But I can read. And I can see. And—brace yourself—I even have times when I can think.

And therein the problem lies: this agonizing process of thought.

To think things through can be painful. To reason things out can paint you into a corner, for to stay inside the lines of logic—non-contradiction, whatever is, is—can take you outside of your comfort zone.

But if that is true, then this is ...

But that would mean I've been ...

Then what I've been taught is ...

Disquieting thoughts? Perhaps. One thing is certain. If systematic *thought* impels such deductions, it's little wonder *logic* has fallen from favour. Or, if not fallen, then selectively held—to where rejection of logic becomes the logical choice.

With life but a venture in madness, why bother with principles of reason?

If electrons can hold two positions at once, then why can't I do the same?

So much information: Who could ever dig through it?

Good questions.

But what if all answers aren't buried? All diamonds aren't down in the mine. Sometimes all it takes is the kick of a boot—a gem lying there for the picking.

Maybe some answers can be found that way, too.

Lanterns on the Nashwaak is a collection of stories inspired by the people, the homesteads, the ecologies, and the landscapes of the Nashwaak River valley.

But the chapters hold more than narratives. They contain seeds of thought — existential meditations; reflections engendered from walking my dogs in the field off Sweeney Road, the same ground that grew *Apples on the Nashwaak* (2019 Chapel Street Editions). Serene, secluded, a land of introspection — a place impossible to walk and not think.

And … when I do think?

Well, whatever *thought* is, it has the power to push. For when I stand there enamored with the natural world, I'm ever constrained to go deeper — forced to contemplate worlds philosophical, to reflect upon realms scientific.

But pitfalls crater the pathway of consecutive thought. While following straight lines may seem the sensible course, spiders spin the same into webs. And with curved lines cocooning their prey? You can think yourself in circles trying to square reason with the claims deduced from the evidence.

Careful. I said 'the *claims* deduced from the evidence' — not the evidence itself. Yes, the facts stand. But the assertions made, the inferences drawn, this is where logic lays bare the inconsistencies. For between what I can glean for myself, and what I'm told are the truths the evidence produces, there, manured with the dearth of reason, lies a now-deserted land filled with how-can-it-bes.

Yet it's a plot that must be explored.

So take heed, fellow-hiker. Rocky trail ahead. Not a few places will cause us to stumble.

But we will walk together.

We will come through as friends.

And maybe with a handful of a diamonds.

Introduction

Two women stand on a seashore:

glowering

clouds,

firebolts flash through phosphorus

smoke,

froth veins the face of the water.

One woman, an author; the other, a poet. Give each a notebook and pen. What will appear on the page?

The poet suffers the blows: slap on the face, punch to the gut, the poet is hit with the *now*. Wounded. Bruised. The colours, the sounds, the unfathomed depths — the grip within her chest, constricting, the poet cries out, *I must reduce these impressions to verse.*

The author slips into fantasies. A bite of the lip, and a scratch of the chin, she contemplates all that could be: that boat on the horizon caught unawares; those rocks, and the run of the tide. The author charts a path to some distant shore. She whispers as she scribbles her thoughts: *Someday I'll expand these into a story.*

One ocean storm, two witnesses — the poet going down for the final time; the author sifting through Life Guard resumes.

Both impulses are used in story. A shift of the fulcrum can better the balance. If held too long, too sharp a focus can render too narrow a view.

To be entranced by a sparkle in a dewdrop is to miss the gift of rain on a field of thirsty corn: Farmer Brown's cash crop, his daughter's university tuition, her fourth year of medical research, immunotherapy trials for pancreatic cancer; a chance her mom lives to see her first grandchild — there's a story in the panoramic as well. A rain spread over five-hundred acres — fully as romantic as a pinprick of sun.

Prose, and poem — the someday and the now. Blended or standing alone, both comprise story.

You'll find both here.

Note to reader: Numbers are written in multiple formats — some for contrast or emphasis, and some for rhythm or flow. You'll find every mode but consistency.

Reflection 1 – Mind

… this field … with my dogs …
it's close to the house, and
quiet –
sadly
quiet, birds down by half:
cats in the grass,
reflections in glass,
viruses nested in feeders;
whole time zones of timber
razed,
plundered by aliens:
boom arms, cable
necks, rotating
heads –
war-of-the-worlds harvesters,
deaf to the screech of the killdeer,
cleave the last contiguous stand.

Above the trampled swaths,
pewees tread air,
brood the devastation from these
lost-nest monsters.

But we must get these trees to the mill,
planed lumber to port; stacked

decks — teetered
freighters fast-bound for Tanzania.
Board feet bandied for cloves;
anything to flavour our hams,
to numb the pain
of aching teeth. Can
cloves be pushed in
like pins
acupuncture — driven
deep, both to spice and
to stupefy
the vapid existence
of the human soul?

Lucky dogs.
Not forced to think.

Man. Woman. Humankind.

Each person, a physical being: skin, bone, muscle, and blood. A framework you can see and touch. No mind required. A brain, yes — that electrified mass couched in a skull — a brain's needed to run the show, but 'mind', initially, is a camouflaged component.

A baby slips into the world. No need to think — to tell her heart to beat, her lungs to breathe, her bladder to empty at three in the morning. *Brain* has it under control. Automatic, autonomic, that visible clump — a three-quarter-pound bundle of axons and dendrites — oversees the body's physiology. Keeps the parts moving together.

But the child grows: begins to show qualities uniquely her own. Quirks and characteristics emerge, identifying her as ... her. And keep watching, Mom, for that little thing known as a *stubborn streak* — when you'll be forced to say, "That girl has a *mind* of her own."

Emotions, habits, eccentricities: stirred into the mix, seasoned with foibles — traits coalescing into personhood.

And though we pierce the mysterious here, these attributes of identity must emanate from a secret place — a nook not found within tangible brain, but housed in invisible mind.

Reflection 2 – Wonder

Six dogs run a thread
through these last twenty years, but
never have we come here at night;
maybe twice … eyes turned
to the star-stream,
to that spilt-milk spiral-in-spin –
for what of that pair
of devil-red eyes, that
riveted stare from the grove?
(Bravado in whisper:
I tell Lil and Boo not to fear.)
    I know owls of orange eyes
    raven at dusk,
    owls of eyes yellow, at dawn.
    But these? Slit
    jaspers
    imbedded in velvet noir?
    Alligator? Yes,
    but there should be no species of alligator here …
    feeding,
    at night,
    with us,
    their prey!

There is no new breed
of which I'm aware – though only last month
I let lapse
my yearly subscription to
"New Frontiers
in Mutant Reptilian Ferocity".

Hear the barred owl?
Not bard, as in Shakespeare, not
barred – verboten,
but, barred as on windows;
imprisoned:
her wisdom, written in feathers:
Flight – that hope of escape
from Earth, or
from bondage – leaves both bird and man
with bars down their backs, for
however high or distant the glide,
all wings come fitted with feet.
And no man can
ever

break free from
himself.

Eyes: gelatinous globes hot-wired to the brain. Orbs to optics. Impulses turned into sight.

But the eyes show more than what we can see; they transport to the sublime.

The three of us. Together. Those night-walks in the field? I lift my eyes to the stars. They turn their heads with me — they follow my gaze, expecting a bird on a limb. (And though dogs aren't as colourblind as previously thought, a full-spectrum analysis isn't needed: jays — blue or grey — they hate 'em both.)

But here we three stand, faces to the sky. They see what I see — see all that I see; see more than I see, at night. Yet my eyes go further than theirs do:

How many stars are up there?

How did they get there?

How deep, the heavens?

Dogs don't think such things. Whales or dolphins don't either. Or gorillas or chimps or squids or crows.

It's not that they've 'lost the wonder of it all'. They never had it in the first place.

Physical brain.

Invisible mind.

But humankind goes deeper yet.

Reflection 3 – Conscience

I'd been promising for weeks. "Yes, guys, I know. Snow's gone. Our field's green again. Time for a picnic and a good long run."

I make three chicken sandwiches: two, just chicken; one, with chicken and mayo and pepper and pickles (some days it don't pay to be a dog), and olives and croutons and five strips of back bacon, crispy.

All systems go. We're off – dashing to the field, and barking all the way.

I lay out the blanket, and we get down to it. A bowl of cool water, a thermos of tea, napkins for plates – side courses, three.

Boo chows down and runs to the orchard. He says that partridge he started last fall is wanting to pick up the chase. From the flutters and the growls, sounds like it might be. Lil finishes up, then follows.

Now, to a sip of tea.

I slip my hands into the canvas lunch bag (both hands – olives are expensive). I'm about to lift my ... layered composition, when I hear a yelp from the trees. Not a "Help! I've stepped on a bear trap" yelp; this, more a "Master, your presence is required."

Slurps and drools now on hold, I return the sandwich to the bag, then head off to check on the dogs.

Bad move. I enter the orchard just in time ... to see Mister Fox running off with my meal.

"Stop. Thief!" Just my luck. Five million foxes on the eastern seaboard, and I get the only one that's deaf. Doesn't look back. Heads to the pines at the end of the field. He drops the bag, shoves in a snout – his back lurches with every gulp. Lays down when he's done. Licks his paws in the sun. Yawns. Stretches. How could he? The guy just stole my lunch, and he lays there, all smiles?

Next day, there's a knock at my door. "Hey. Someone's here, guys. I bet it's that thieving fox – come to confess. He's having trouble falling asleep."

I stole a *Batman* comic once: #142- *Ruler of the Bewitched Valley*. Wren's Drug Store. 1961. Quite a transaction. Though I came out ahead, I ended up the loser: bought a life-long memory with a dime I never had.

Why do we feel so bad sometimes?

And 'Why?' is only half of the puzzle. The more difficult question is 'How?' How can a physical life-form FEEL bad? Not from physical pain—nerves pounding from a hammer to the thumb—that type of pain's understandable. But the kind that melts into the emotions, that flows into every corner of the mind—how so, this type of pain, if all we are is a physical being?

Whether we like it or not, there's something within us called *conscience* – a mouthy little guy who knows all about us.

How did this incorporeal 'person' get inside me—me, but a mass of muscle and bone? Where did *he* come from? And why is he never satisfied, never at peace? And how does he know the second *I* awake? Couldn't he stay in bed just another half hour—give me a tranquil head-start to the day?

Hey, you in there? Leave me alone. If you won't go to sleep, then, here, let me help: this drink, this pill; try one, try both. Must be something I can swallow to shut you up!

Too bad about dogs—can't look in wonder at the stars.

Still, it must be a treat: no little guy running around in there.

Reflection 4 – Origin

Sunset is lovely at this time of day,
but I wish it would befall hours earlier.
Maybe high-noon, or,
just before tea,
either one of these wishes would
work well with me.
Sunset at sunset just ruins my day.
It's high-time
we set
things right.
So, you with your
Rolex, and
me with my Black Dial Longines –
we'll show that Ol' Sol who's boss.
Synchronize watches in … three … two … Wait!
Do you not hear that yellow
guy-in-the-sky
making light of our timely concerns?
Talk about a flair for solar wind!
Sun says, "No matter
the make, the
style
– whether works of
titanium or tin –

*you can wind any watch, but
after a while,
time's lost
in crystal and cesium.
When it comes to chronology,
I own it whole;
your watches are only time-
pieces."*

*So ... Sun says he'll keep setting
at sunset —
precisely, exactly, at
sunset,
and sunset will stay where it is —
always at the end of the day.*

*He makes a core point:
He hasn't lost a second in all these years.*

*Lost time's only found
in the measurements of man.*

So ... *In the beginning?*

How many years ago was that? Fourteen billion? Eighty-five million? Two hundred thousand, or ten of the same? You weren't there. I wasn't there. Neither was Darwin, nor Moses. No onlookers. No beholders. And we've certainly no eyewitness-verified facts.

Think of that. In the however-long history of humankind, as we stand on that timeline today, we have, at most, a one-hundred-and-twenty-year span of seen, verifiable, testimony — where someone can say, "I was there, and here's what I saw."

Maybe one hundred years of verbal affirmation.

To learn of the days preceding, we have words, symbols, passed down on parchment, paper, or clay — what another has written, left for us to read, to interpret.

And therein's the rub: interpretation. Should be a simple enough process — read, then move on to objective science: observe, measure, verify, and repeat. Make a declaration based on the evidence.

Surely we'll arrive at a common conclusion.

Unless — as with my dogs — we're all just a bit colourblind.

Reflection 5 – Life

I'm sorry, sir. That particular soup has fallen out of flavour

Beef broth in beakers,
glass goosenecks.
Hard boiled. All
life, destroyed.

In half the carafes –
necks broken,
contents exposed;
clouds of bacterial contaminates,
migrating microbes;
bugs on the backs of dust beams,
slid in on shafts of light.

But from the half kept sealed –
the goosenecks, sound
the cry of a thousand tongues:
the broth (and canon)
clear:
"Spontaneous generation a chimera,"
wrote Louis.
"Gaia cannot
bring life from stone. Life
comes from life
alone."

But all tastes change
with time,
more time, more
time pushed ad infinitum,
beyond the bound of beholding.

Louis' soup's
now off the menu:

recipe, tossed —
science,
trodden.

The most basic question of all: How did that first life appear?

Watch your terms. Not, how did that first life *evolve*, but, how did that first life *appear*? A question of pre-biology — the first living cell having necessarily arisen from non-living elements.

No matter how many steps you walk back through time, first-*life* had to have come from *non-life* material. But this is a postulate already (and many times since) disproven. A slurp of Louis' soup: spontaneous generation, a fantasy. Life can only come from life. q.v. Redi, Pasteur, et al.

So we're told if we went back far enough, we'd enter an epoch where life-from-non-life occurred all the ... time, to an age of disparate conditions:

Pasteur's experiment only proved that spontaneous generation couldn't happen in the conditions on the earth in his time. Had he done the experiment 5 billion years earlier, he would have gotten a different result.

I see. So a pinch of non-uniformitarianism—where the present *isn't* the key to the past—is permissible on occasion?

All sounds *quite* scientific.

But tell you what: Forget the soup. Just gimme a burger. Brontosaurus is fine.

Better make it two. My friend must be starved—the monkey there typing out Shakespeare.

Reflection 6 – Simple cells

"Goin' in the truck?"

That'll start the spins and the whines – pawing the glass in the front door window.

The field's a great place to run, but I do have to be careful. I sound the horn before we exit the truck. And I always carry a cowbell. Bears, coyotes – I've seen them both; thus far, at a distance, and before the dogs have caught wind.

But there are other things up here, too: other dangers – maybe without cuspids, but with every bit of a grip.

The last few years have seen an upsurge in deer ticks, carriers of Lyme disease. Can be a killer for dogs. (Not so nice on their humanfolk, either.)

I check Lil and Boo after every run. And I always carry a tick spoon. Works great – on all of us. One constant-pressured drag from my special spoon and ... off they come: head, jaws, scolex, molars, fins, fangs – whatever other rigging those bloody things got going.

Hate ticks? Yes. But you've got to love the word that describes how they hunt. A great word. Have you heard it?

When they crawl to the tips of the grass blades, when they're hanging on by their rear legs waiting for a passing shin – when they get right down to hunting for prey – this pose is known as questing. *That's what they're doing:* questing. *Terrific terminology, but I do see a problem: Should the first lunge prove unsuccessful, then the second one would mean they're* re-questing.

And my answer will always be 'No'.

Take a tick attack – a timely concern, as we've suffered four assaults this year, along with four press-n-drags of my spoon. The last of these surgeries proved interesting.

Back from our run, I did our usual check. *Oh-oh. Found one.* I removed the tick, but ... *Hey! What's this crawling on the deer tick's back?*

I looked at it under a magnifying glass. It was a scydosella beetle on the back of the tick. *But what's this crawling on the back of the beetle?*

Dug out the microscope. Found a mite on the back of the beetle on the back of the tick. *But what's this on the back of the mite?*

I increased the magnification. *Look at this: a protozoa on the back of the mite on the back of the beetle on the back of the tick. But what's this moving within the protozoa?*

Proto-zoa. First animal. Little animal. Single-celled. (That's 'single', not 'simple'. Seems only humans have 'simple' cells. Brain cells, blood cells, skin cells, fat cells: 200 different types, 55 trillion in total, last count.)

Here's a partial list of our simple cell's constituent parts:

* Cell Membrane – the outer skin. It's a 'smart' skin—selects what's allowed in and out of the cell.

* Cytoplasm – the jelly-like substance within the cell; jelly, but with bones, of sorts: a cytoskeleton—a protein framework—keeps the cell from collapsing.

* Nucleus - the cell's brain, if you will; tells the cell when to divide. Contains most of the cell's genetic material.

* Nucleotides – the building blocks of DNA. There are more than six billion nucleotides in every human cell—three billion base pairs, the rungs on the DNA ladder that deal with information coding and storage.

* Chromosomes – the thread-like DNA-wraps located in the nucleus. Twenty-three pairs of various lengths. Chromosomes are composed of genes and the spaces between them.

* Genes – regions of DNA, composed of 1000-50,000 nucleotides, the letters that write our protein code.

* RNA – deals with information messaging and transport; the intermediary between DNA and protein production.

* Mitochondria – the cell's power plants; convert sugar into energy. Contain a small amount of DNA passed down through the maternal line.

* Ribosomes – processes proteins from the codes in the messenger RNA (mRNA).

Parts of the simple cell — individual structures, working in harmony to perform respiration, elimination, reproduction. A library of data — 1.5 gigabytes in every cell.

And though our cells are continually dying, the body's ever calling up the reserves. With red blood cells alone, 2,500,000 are made every second, 200,000,000,000 per day.

46 chromosomes, 20,000+ genes, 100,000 different proteins, 3 billion base pairs of nucleotides — in every cell, and all of this, information: more information than minds can compute.

Cells. Tiny cities.

Simple, we call them.

Okay.

So. Ya wanna buy a bridge?

Reflection 7 – DNA

Monday afternoon in my field.

They know something's wrong; they
sense I am galled.
Sharp readers, dogs.

Now …
I can understand the odd empty –
bounced from a four-
wheeler, or
dropped by a hunter, who, caught
unawares in a temperature
spike (shot up from buck-fever),
disgorges his Pepsi while
drawing a bead; that
warm
syrup
gumming up the gun's
magazine;
ten sticky fingers – all
factors which trigger his
missing the shot;
a left-behind empty in a case
such as this?
I can see it.
I'll pick up the can.

But, these —
pustules
smeared and
spreading? These
lesions of litter strewn
wholesale?
From the edyoucated, too,
I'm told —
a graduation party gone wrong:
twenty-five revellers, two hundred textbooks
— not a one with a spine —
50,000 pages torn to the wind:
whitecaps,
pillowcases
pressed into paintbrush —
til a cross-current gust sets them free.
Breakouts in backflip,
no numbers
sequential —
all of that knowledge, not
spread, but
dispersed

information
lost to the wind.

Pick a letter.
Now, pick three more.

Pick any four letters you like, as long as they're A, T, C G: Adenine, Thymine, Cytosine, and Guanine—the nitrogenous bases of your DNA.

Each letter takes on a sugar and a phosphate—becomes a nucleotide. C bonds to G, A bonds to T—links in a chain of 3,000,000,000 base pairs, rungs in the double-helix ladder. When a certain protein is needed, the hydrogen bonds break, the ladder unzips, and three adjoining nucleotides encode for a specific amino acid, the building blocks of proteins.

And though we're talking 'tiny' here, six feet of DNA is contained in every cell. Multiply this by the body's 55,000,000,000,000 cells—you get 300 round-trips to the sun.

Ever-increasing complexity. We now know that in any stretch of DNA, there are overlapping codes on the run—up to 7 codes being read simultaneously: codes running backwards and forwards, and with different starting and stopping points.

All that molecular machinery. A trillion-plus bits of stored information.

Hopefully, nothing goes wrong.

Something always goes wrong.

Reflection 8 – Mutations

Bridget Evans was the oldest in a herd of eight siblings; the oldest and the shortest – stood four-foot-eleven.

Her share of the chores was the milking. Though only five head, at six every morning, at six every night, she was found ever-faithful in her two timed squeezings.

But it's what Bridget did before the milking that would make her a girl of renown, that sort sung about when children skip rope.

In the pre-dawn hours, Bridget taught herself typing and shorthand. Though no-one was disturbed by her squiggling, the click-clack-ziiiiiing of the typewriter forced her into the barn. The cows didn't care. They even gave a bump in production: misheard the keys as crickets.

On the day she hit ninety words-a-minute, Bridget tore out of the Zionville flats. Landed a job at Brown's Insurance on the corner of York and York. (Fredericton was much smaller then.)

Her first week there, she ran into Tammy. The two had lots in common: both secretaries, both from the farm; both shared an interest in reading and writing, and they both shared an interest in Paul. It was here Bridget feared her rival held the pole position – Tammy, a towering five-ten.

Something had to be done.

Bridget started coming in early – was at her desk a full hour before work. She told Tammy she was writing a story that would shake the foundations of the literary world. She typed two pages every day: day after day after day. Both folder and plot began to … like what happens when you simmer tapioca.

Tammy had to know what was in that file. She'd do anything to have Bridget's story in her greasy, green-eyed hands: pay a thief, threaten harm; offer up a gift sacrificial. Paul would make an excellent lamb.

She made excuses for not going to lunch — insisted Paul and Bridget go without her. Even with the office pool's staggered lunchtimes, Tammy calculated she'd have ten minutes alone — time enough to piecemeal Bridget's file, reverse engineer it via shorthand. Once the backlog was cleared, two pilferings a week should keep her abreast.

Though Bridget was thrilled to be dining with Paul, she soon named these chapters **Noon Hours to Nowhere.** *Paul's every topic was Tammy-Tam-Tam. Bridget was losing her appetite: every item on the menu, sticking in her craw (though she found herself developing a real taste for blood).*

One night (in gloves), she broke into Paul's apartment — Paul off to visit his mommy. Bridget laced a jug of lemonade with strychnine (every milkmaid knows how to keep rats from the barn); hid one of Tammy's earrings in the couch.

When Paul was three days missing, his cousins came knocking. They found him dead on the bathroom floor, knotted into a pretzel.

The police began a murder investigation. Bridget overheard the phone call — Sergeant Miller telling her boss they'd be questioning the office staff the next morning. At closing time that night, Bridget removed the pages from her story file. She returned the next morning, ten minutes earlier than usual.

The troopers stormed in at 8:30. They found a strychnine container and a matching earring hidden in Tammy's desk. And … one of the detectives knew shorthand.

He began to reference some choice cuts from Tammy's novel: close-sounding names, locations; street layouts identical to where Paul lived, the same apartment number — all written in Tammy's own hand. Tammy was arrested — under protest, of course. While she admitted to writing the story, she said it wasn't hers — just a copy. Told the officers to check Bridget's desk: third drawer down.

Bridget fell into her chair. Said she'd never been to Paul's apartment — didn't even know where he lived. The sergeant wheeled her back from the desk, opened the drawer, and pulled out a folder.

He read the title-page: "Chloe Carmen and the Horse Race at the County Fair" – a snuggly fifty-pager for eight-to-ten-year-olds.

* * *

Paul, murdered; Tammy, sentenced to life; Bridget's children's story published nationwide – in every primary school in the country. "Local Author Hits Big Time", the headlines read.

Three days before Bridget was to be given the key to the city, Paul's niece in Calgary reread her copy of Chloe (only for the umpteenth time). Something set her to chewing her pencil. Something about the horses' names. She checked her scrapbook of newspaper clippings: showed her mother what she'd found.

Sergeant Miller made another phone call. Could Bridget come down to the station? Just a few questions he'd like to clear up. Bridget said she'd be there directly. Ended up taking a detour ... over the Carleton Street Bridge.

Bellha and Tryxee: nice names for horses. Weird spellings, though. Some letters doubled; some, reversed, but exactly as written on the back of the picture Paul kept on top of his fridge – the two horses he'd had as a child. The picture that fell to the floor; the names Bridget read as she stirred the lemonade.

And the children skip rope as they sing:

Midget Bridget

Pretzel Paul

Who's behind the jailhouse wall?

Toxic Tammy,

that be she.

How many lifetimes til she's free?

One, two, three, four, five, six, seven ...

Mutations: the 'somethings' that go wrong.

When you're working with long lines of letters, typos are bound to occur. And they do—from individual bases, to chunks of chromosomes involving any number of genes.

There are different types of mutations:

* Inherited – faults gifted through your family line.

* New – defects that started with you (one hundred mutations per person per generation).

* Acquired – mutations picked up from your environment: pollution, UV-radiation, chemical ingestions, carcinogens, tobacco, preservatives, dope, inhalations of car fumes, nuclear bombs. Your choice. All of these factors (and likely many more) can cause changes, both in your genetic code, and in the expression of the genes themselves (Epigenetics).

A. T. C. G. Letters get omitted, inserted, inverted; duplicated and deleted. And, while Hollywood likes to narrow the focus, the truth is, every one of us is a mutant.

No big deal. What's a missing letter or two?

Sorry. It's more serious than that. Mutations add up. They're cumulative. Like chewing on paint chips, we're laden with lead. And getting heavier all the time.

You carry your own defects, your parents' defects; your grandparents', and their parents', too.

Generations of mutations: ours, and ours to pass on.

All humankind: sick and getting sicker.

Reflection 9 – Speciation

Ellie Johnston had been gone for twenty-two years. Twenty-two years, three months and four days if you were to ask her husband Henry.

The Johnston's had no children – no-one to take over the farm; a growing burden to Henry – alone, 87, and sicker than he showed.

His grief wasn't born from the farm itself – the house, the buildings, or even the land: "Take them all if you want." Henry's worries sprang from his concern for Charlie, his horse.

On the day Ellie died, visitors found a foal in the barn. Proved a real life-saver for Henry. Twenty-two years of companionship; he got over every hurdle with a friend.

But Henry knew he'd be leaving soon. Who'd be there to look after Charlie? And not just to feed and water, but to keep a keen watchout for coyotes.

"Two boxes of 30-30s, young feller," Henry told the clerk at the Taymouth General store. "Used up my ammo shootin' that wolf. But I got him sure. Won't be no trouble to Charlie no more."

"Yes, sir." The clerk reached for the shelf. "Two boxes. 30-30s. But did you say 'wolf'?"

"I did. Eastern Timber. Didn't want to do it; Lord knows I didn't. But I couldn't have him spookin' my Charlie. Too bad, too. Just the second wolf I've seen since McKenzie King took us to war."

Small world, to be sure. Turns out that young feller's uncle ran one of Fredericton's papers: the Capital Free Press. The word soon circled. Teased in type and tongue, Henry now the joke of the Nashwaak.

"There hasn't been a timber wolf round these parts in more than a hundred years."

"The old guy is losing his mind. Eyesight too. He might've winged a fox. A coyote. I'll give him that. But a wolf? Long gone."

* * *

It was a warm evening in May when the cub reporter stepped onto Henry's front porch. Henry had seen him coming.

"Good evening, Mr. Johnston. My name is Andrew, and ... I hope you're out of shells."

Henry smiled as he patted his rifle. "Nope. She's loaded. But have a seat. I'm just keepin' an eye on Charlie over there."

Andrew turned to the pasture. "So that's Charlie. Nice horse. Any more ... wolves around?"

"Nope. Guess one's all there was."

Andrew pulled a notebook from his pocket. Slid the pen from the cover. "Mr. Johnston, I want to say I'm sorry for the way the paper, my paper ... has caused people to — "

"Oh, pay it no mind, Andrew. We surely don't. Ellie and me, we let those things slide."

"Ellie?"

"My wife. We sit out here most nights — on the nights we can; around the stove when the cold has Charlie in the barn. We talk it all out, talk everything over. We don't blame no one for laughin'."

"But, Mr. Johnston, I was told ... sorry, but I was told your wife died years — "

"Twenty-two years, seven months, and four days — same days gone as Charlie is old."

"So your wife didn't — "

"Now you listen here. I'll say it straight up, you bein' with the paper and all. This old man knows his wife's gone; knows she ain't here sittin' where a body can see her — I know it. Don't mean I can't talk to her, does it? Don't mean I don't know what she thinks about things, or how she wants things done."

Andrew looked up, his eyes asking permission.

"Go ahead. You can write that down — me talkin' to someone who ain't here. It's okay. Young people today. Who knows about love *anymore?" Henry turned his eyes to the paddock. "Maybe someday you'll find someone like Ellie is."*

"I'd like to hope for as much, Mr. Johnston."

Henry got up and stood in the doorway. "Here. Come through. Take a look. The fireplace."

Andrew stepped by him, crossed the floor to the chimney. There, pinned on the brick above the mantle, hung Henry's wolf. "See that?" Henry pointed from his stance at the still-opened door. "The reddish-brown muzzle, the black on the back? She's a Timber alright enough."

Andrew fingered a clump of fur. "The Eastern Timber? I don't know much about these things, Mr. Johnston, but why haven't you shown anyone this ... hide, this pelt – whatever it is? Shown someone. Could've saved yourself some mockery."

"Maybe." Henry kept his eyes on the pasture. "And maybe not. Even with evidence starin' you in the face, takes bravery to admit what you don't want to see. 'Sides, it wouldn't've mattered what it was: moose, wolf, bear – I wouldn't let nothin' hurt Charlie. Nope. It's just me and Ellie and one knowin' horse. We've talked it all over. And we've settled on lettin' the whole world go by."

Here's the thing about purebred breeds: The more they're selected down for a particular trait, the more they lose their genetic diversity. The wolf has given us 350 breeds of dog — all variants from within one kind. A great number of species — speciation — and every one down from the wolf.

These facts testify to:

* the initial richness of genetic diversity,

* the subsequent loss of genetic diversity — the fact that you can't get back home (try breeding a purebred back to a wolf), and,

* no matter how many species, or variant breeds, a dog is a dog is a dog.

Again, watch the terms.

Genetic variation? Yes.

New species? Certainly.

But, evolution—climbing from molecule to man? No. Moving ever higher in the so-called 'tree of life'? No. New information being formed, enabling jumps from one kind to another? No. That's not what this is. Nothing here's moving up, up, and away. Here, information is lost. The only direction is down.

Speciation: traits selected from within a genetically-rich kind—bred to produce a particular breed of dog.

Listen. Hear that? There's a new breed barking right now.

Labradoodle? Morkie? Puggle?

New. Cute. But they're all dogs from dogs.

All from within the kind.

Reflection 10 – Second Law of Thermodynamics

Kinnard Gallagher hated the river – this stretch of river, so far from town where different folk lived different lives.

Second generation, fifth child in ten. You'd think fifth in ten would've centred him, but there's no middle child in an even-numbered brood. Four siblings ahead, five behind, Kinnard was left scratching in a play for either side. Too young to be in charge, too old to be babied, and shorter than others his age, his best times were the hours with Reddy and Jack – off with the dogs on dog-day afternoons, on never-ending runs to never-quite-reached destinations.

The dogs taught him tricks, or tried to – poor Kinnard, not a top student. Couldn't catch a rabbit. Never learned to sit, but could chase himself in circles with the best of them. A study in sandpaper and silk: a small, square man in a big, round world.

But a world he was constrained to see.

The fall of 1860. He was painting the roadside face of the barn when he heard a rolling thunderhead tell him it was time. He shook his mother's hand. Shouldered his case. Walked the ten miles to town. Took the paddle steamer on to Saint John.

"Where to now? It'll take most my cash to sail to Boston." Casting the smell of soil upon the open waters, he emptied his pouch at the wicket.

He was in Beantown a week before the celebrations – the city welcoming the Prince of Wales, the future King Edward VII. Kinnard ploughed through the crowd – had to press royal flesh, and did so ... just as the camera flashed. The newspaper spread them all over New England: the Farmer and the Prince – Kinnard turned celebrity, forced to shake half the hands in the city; the other half raising their glasses.

But fame brings her burdens: will choke any man who swallows enough of himself. Had prestige, and the paper's name, The Flag of our Union, weighted the spin of the wheel; set Kinnard on the road to pay-back; sparked a smoldering obligation to his newly-adopted land?

He quit his job on Long Wharf. Joined the Union army – the 20th Massachusetts Volunteer Regiment. Who could've known the play of fate: This man who shook the hand of England's highest prince – pictured from Caribou clear to Long Island – now fighting alongside the grandsons of Paul Revere?

Kinnard fought at Ball's Bluff. Fought at Antietam. Took a shot to the head on the field of Gettysburg. Friendly fire? No-one ever said, but the surgeon left the slug in Kinnard's skull. Blessed him with a medical discharge, and a pension of eight bucks a month.

He hobbled back to the farm in April, '64. Shouted curses at the river. Shot siblings Three, Four, and Six before Seven could grab the gun.

The judge ordered a scaffold built – not in the jail yard, but out on the street. Scheduled the hanging for Monday, May 23rd.

Said there'd be a huge crowd on Victoria Day.

My roof is full of holes. Only twenty years into twenty-five-year shingles, but my roof is full of holes.

Time's a killer: wrinkles skin, thins hair, yellows teeth. Softens memories as it hardens the arteries.

Every*thing* every*where*: wearing down, wearing thin, wearing out. Going stale, or losing its fizz. Checked your *Best Before* date? Ready to be tossed?

I'd love to see a sign at the bakery that read: "3-day-old bread; FRESHER now than the day it was made." (Who cares whether crystallizing starch molecules have put some order into the system—the bread's gone stale! Gimme a fresh, disordered loaf.)

Or, this sign at the garage: "We guarantee our tire treads will be an inch thicker after fifty thousand miles."

Two signs you'll never see.

Systems left to themselves have a tendency towards disorder—a disorder that increases over time. In any closed system, there's a decrease in workable energy, a slide to thermodynamic equilibrium, that state where no more change will occur.

Now, while food breaking down in digestion, or the sun losing energy to the Earth, are beneficial processes, still, energy is expended; the energy available for work, constantly decreasing.

If you don't re-paint, re-shingle, or re-apply cream to the face, then stand back and watch your colours fade, your shingles thin, your face ripen from plum to prune.

Nothing just sits there. Entropy (disorder) increases over time.

It ain't just your teeth.

The whole universe is in decay.

Reflection 11 – Genetic Entropy

Hugh Thomas Good lived bound to the earth, imprisoned by friable soil. Every day full, and every day long, he loved every facet of farming.

But tares grow too with the grain. Wild onion with wheat, green foxtail with corn, rivals root in mutual beds – plant against plant in a garden row, thought versus thought in the mind field. It's here where love and hate bear their singular, bitter, fruit – a familiar harvest to Hugh.

Those days that follow the last melt of snow – rebirth of birdsong; pussy willow, pin cherry, blossoms now spent. Warming winds whispering promise: green leaf and blade at the door. But "Hold on, Hugh. Not yet." To plough now would chance compaction – would compress the soil to concrete, would bake a turned sod to clay. These weeks of waiting with their press for patience, this 'to every thing there's a season' time – this was Hugh's time to hate. Life's spirit on hold – **Terra Interregnum** *– spring's interlude with its seasonal depression.*

He felt the first twinges in April of '32. Not the stitches sown from a long winter's layover – these, deeper: sprung from a sense of slipping time, of being used up. Hugh never said a word. He kept up a front for three years running, but everyone could see he was fading – the drag of those not-quite-yet-springs weighing heavier upon him each year.

No problem in swearing his doctor to silence – the doc didn't know what it was. He tried potions, concoctions, elixirs, none with any measurable success. It's not easy selecting a cure when you don't know the source of the malady. And the names of the ailments didn't help: black dog (depression); green fever (anemia); falling sickness (epilepsy); the palsy (tremors). Medical analyses of yesteryear – well before these days of reading 5000 diseases in the genes.

"So what's my problem, Doc? Something specific, or general wear and tear?"

"Hugh," the doctor sighed, "I really don't know. Sometimes a man takes sudden sick. Other times his bones will cry out for sleep — simply worn out, threadbare from years of decline."

Entropy goes deeper than fading paint.

Remember Grandfather's eight-day clock? It needed attention every Saturday night. You had to wind it up 'cause the clock was winding down. While the mainspring put energy into the system — turned the hands, allowed the cuckoo to sing — the clock used up the energy available for work: ran it down, down, to Slow, then Stop. This is the direction of every system, of every natural process left to itself. The law is universal.

And deep.

For entropy's not only 'out there'. It's 'in here', too. We're all winding down, going into disorder. It's in the very blueprint of life.

Our cellular machinery's under constant attack — enemies bombarding our lines of code: insertions, deletions, substitutions — a whole host of typographical corruptions.

Compacted as lead, diffuse as gossamer, mutations are killing us all. And their numbers are ever-increasing. When you add the hundred mutations per person per generation to the mutational load inherited from Mom and Dad, humankind is into a build-up (a breakdown, better said). And while most mutations are not immediately life-threatening, each one reduces the effectiveness of proteins. The best that can be said: "Hey, they're only slightly deleterious." Granted. But with the body being hit with a million mutations per second, clandestine under-workings are herding us over the hill.

What shall we do?

What can we do?

Eugenics? Sure, Adolph, go ahead. But tell me again, who gets to decide who's superior?

Cloning? Sorry. Problems aplenty: high rates of deformities, and clones come already aged (and please don't think 'wine and cheese'), resulting in premature deaths. We don't yet know why, but clones are inferior to their 'parents'. And the clone — you guessed it — still carries its own mutations.

Then why not remove the bad genes? Good luck. While true for some, most diseases are not of the one-bad-apple-gene variety. And with most mutations being near-neutral negatives, they are extremely difficult to select out. The genome doesn't exist in strings of single, isolated nucleotides, but in 'blocks' or 'chunks' of interacting, overlapping codes; as many as 20,000-40,000 nucleotides — carrying multiple messages — info going forward, backward, and even in 3D. While gene-editing technologies such as CRISPR hold promise, significant challenges remain: accuracy in DNA slicing — off-target editing — not the least of the concerns.

The human race: Everybody's in it, and the humans are losing. We're already deep into genomic degeneration, well on the road (rather, 'sick on the road') to mutational meltdown.

Hugh's doctor was right.

A man can be brought down in many ways: the load on his back, the load on his mind. Soon to be the load on his genes.

Reflection 12 – Evolution

It's snowing on the Nashwaak today:
not a blizzardy snow, not a
wet snow
  – stuff that balls sticky on dogs' legs –
not coarse, crumbly pellets – your
feet grinding salt – not
driving flakes – pin-pricks,
and blinding;
this snow, slow yo-yos
in recoil,
feathers in pendulum arcs,
floaters
adrift
neath an alabaster sun
in the shrouds of an ashen sky –
shy flakes of snow;
hung parachutes, targeted
landings;
will pull every string
to skirt open water, that
quarter-second 'pssst' –
dissolved,
reclaimed:
held promise in a grizzled sky.

The Inuit have words for snow:
snow falling, snow
fallen, snow
best for quick sleds,
each word structured,
as crystals.

But I need the word
for this *snow:*
a word, precise, so
sharp as to slice a
single
flake
into drifts deep enough to enfold me a thousand years.
I need that *word.*

For I cannot depict
a world frosted white

with words that melt at grey.

We use many words like we use the word 'snow' — words so all-encompassing, any application will do.

Old. Love. Normal. Spicy. Evolution.

What? Yes, *evolution* — a snow word if ever there was one. All terms biological go under its blanket. Define the entries in your glossary, then use them however you like.

Evolution: All life forms — both past and present, both animal and plant, extant, extinct — have, by moving through innumerable, incremental changes, developed from one single cell. A molecule-to-

man advancement — the changing from one kind to another kind ... to another kind ... until, e-v-e-n-t-u-a-l-l-y, humankind appeared.

Now, if this is what is meant by *evolution* (and it is), then why is the term used outside of, and even contrary to, its denotation? Why the distance between what the word *means*, and the way the word is *used*? As follows:

Evolution (changing from one kind to another kind, i.e., movement *between* kinds) is used in reference to *adaptation*, a process which relates to changes *within* the kind. It's used when referring to *variation* (again, changes *within* the kind), or, to *speciation* (... *within* the kind), or, to *natural selection* (... *within* the kind).

Why is *evolution* blended into all of these terms? A real mix-up if words are to have any meaning at all. Can only create confusion.

Where's the cry from the wordsmiths? I understand their silence, for there's more here than the tangling of terminologies. It goes deeper. Blending dilutes. Blending pollutes. And, it more than minces the meanings of *words* — it fuses the two types of *science*.

* Operational science — observable, testable, repeatable: the science one can do in the here and the now. As mentioned above, the selective breeding for a particular genetic trait to give a particular species of dog, follows the lines, and the laws, of operational science.

* Historical science — reaches into the past, to where processes and events can no longer be seen, tested, or repeated: to where theories and interpretations of times long gone can't be based upon evidence in hand; to where all that can be held are assumptions: not tangible nuggets mined in the lab, but imaginary fines dredged from the slurry of tailing ponds.

By blending *evolution* (which demands movement between kinds — a process never seen, but assumed to have occurred somewhere in the realm of historical science) — by blending *evolution* with *adaptation, variation, speciation* (all of which allow for movement within kinds — a process readily observed; one which can and does occur due to the depth and richness innate to the genetic code), this fusing the two into one, gives assumptive historical science an air of respectability. Raises its stature. Binds the unseen with observational science.

To use the term *evolution* (between kinds), when all you really mean is that you have finches (within kind) with variations (within kind) in the shape of their beaks (within kind), is to give *evolution* a meaning beyond its definition.

Further, to colour a true, observational change that has come from genetic variation — to colour it in such a way as to give the impression that a species is 'evolving to a higher plane', is, at best, a mishmash of terms; at worst, an attempt at deliberate deception.

Why the union of the two? Because historical science is not science at all. It cannot stand on its own. But fusing the two allows for a common defence: lessens the chances of attack on the basis of reason.

Such a ploy is used all the time: *What idiot would question the science of evolution? Just look at all the breeds of dogs.*

Hear the mix? The conflation? A straw man serving red herring. There's a whole lot of blending going on out there.

Evolution – a snow word. Used to blanket, to bury, in hopes that you don't catch the drift.

Oh. Look. Snow's falling on the Nashwaak today.

Really? I don't want to lose the romance of it all, but, what kind of snow do you mean?

Reflection 13 – Natural Selection

Yesterday dropped down
deep-winter thick:
two feet of snow,
glistering
ice, leaden
earth neath the weight
of a gunmetal sky.

Orchard companions –
apples and deer.
Four browse twig tips, frozen
fruit.
They don't scent the coyote at the top of the ridge,
but
he
sees
them.

Wind shift.
Deer lift heads, then
tails, you
lose – poor yearling runt.

But the coyote won't take the weakest.
The hardiest deer, the
heaviest –

a hoof through the crust.
And another.
She's the one
chosen
as three run free.

Survival of the luckiest —
the fittest, snagged in snow,
a natural selection for a half-starved dog.

The battle's
not always to the strong,
to the fit:

Even the lusty
lose to Lady Luck.

Natural Selection is not a lot of things.

It's not 'evolution'.

It's not new information being created.

Natural selection is the process through which certain already-existing traits prove the most beneficial in a particular environment, and are passed on to the next generation.

Consider our poor frail doe. No, that's not right — our *frail* doe came through just fine. It's the poor *hardy* doe we should pity. The environment she was in — a not-quite-thick-enough fresh sheet of ice resting on two feet of snow — played a major role in her death. How unlucky she was to have been healthy: the hearty one — the fittest — the first to be eaten.

And who's left to carry on the line? Why, the three weaker ones—the lighter escapees, these are the deer that will breed. The 'genetically superior' just got gobbled up.

It's the same for that oak tree over there. Two thousand acorns, but you don't get your trees from the acorns sporting the best genes. You get your best trees—your only trees—from the nuts the squirrels leave behind.

Eaten by coyotes, or eaten by squirrels, it's a predators' world out there.

And the family with the most kids wins.

Reflection 14 – Variation

    Mort Tierney had a problem: Things were too good on the farm. With his cattle, his lumber, and a growing market for produce, his present surroundings – neighbours and riverbanks – had him completely boxed in.

    He needed a different location, a larger parcel of land. And his plan was to grab it on Thursday.

    Thursday was the day William August Bubar sat chewing his pipe at the Taymouth General Store – least he had every week for thirty years running. Mort knew Will owned property in Durham Bridge, and he might be willing to sell. But to broach a purchase while keeping the upper hand would demand a dose of select circumspection. Both men were shrewd negotiators, and both, of very few words.

    Market day arrived. Mort choked the neck of his money bag (his mother was a Campbell), stuffed it deep into his coat, and headed off to battle Will Bubar. Will was there: suited in mail, ensconced on the battlement; crossbow cocked, sights dabbed with spittle.

    Mort glugged a root beer. Pulled his palm down his beard. "Heard you're selling land in Durham."

"Maybe."

"Heard it's mostly rock."

"Yup – same as what diamonds are."

    Oh-oh. Just ten seconds in, and Will had him over a pickle barrel. Mort cleared his throat, his held-in-reserve sweet-talk now drenched in vinegar. "Has to be special land. Five stipulations."

"One?" *demanded Will.*

"Has to have softwood."

"All spruce and fir. Two?"

"Has to have water."

"Got the Nashwaak, two streams, and a spring that never dries. Three?"

"Southern exposure."

"You can tan every day of the year. Four?

"Least fifteen acres."

"Runs twenty. Last?"

Mort slipped his hand into his breast pocket. Cupped his money bag. "Five," he swallowed, "it has to be ... free."

Will picked at a finger. Tore a hangnail off with his teeth. Spit a chunk of something onto the floor. "How much in the bag?"

"Just enough," said Mort. He laid it beside Will's glass.

"Sold," said Will. He snatched the bag and walked to the door. "The lawyer's expecting us at three."

"I'll be there."

"Make sure it's you. Not your twin."

We heard it all the time when we were kids: the uniqueness of snowflakes.

Or maybe Mom and Dad just wanted us to stay outside. "Boys? Of the gazillion snowflakes that are falling right now, only two are exactly alike. Go find the matching pair." (I know they loved us, but my brother, Tom, ended up moving to the Yukon. Haven't seen him since '68. He's been one busy boy up there.)

No two flakes alike. The same can be said for everything. Even 'identical' isn't. Traits can be close, but there are always variations—a fact as plain as that big or bulbous or crooked or pocked or wrinkled or hooked or Greek-gothic nose on your face. No two creatures are exactly alike, human creatures included.

Now, while there's an infinite variety of variations, these modifications come from *within* the already-encoded DNA— modifications from *within* the kind, originating from the code. This is not a particles-to-people, molecule-to-man, progression; no new information is being made. Variations are (dare we say 'simply'?) differences which emerge from that which is already within the kind. Variation says, *Look in my suitcase. Here's another tie-and-sock combination from the million possibilities I've long had packed for the trip.*

This is variation. This is adaptation. This is not evolution — as used in 'one kind moving to another kind'.

Be careful. For you stand on dangerous ground. To mix these terms is to hang yourself.

See where the misnomering takes you. (It's a terrible read, but do press on.)

Consider the differences which occur within any species of bird, for example: variations in colour, in the size and the shape of the beak, in the length of their tail feathers, etc.

The claim is made that these trait variations are examples of ongoing evolution, changes pushing the 'fittest' bird up another rung on the ladder.

Alright. But if this is the meaning of evolution, then be consistent. If you call these variations in one species, evolution, then you must call similar variations in other species by the same term: evolution. If it's *evolution* in birds, it's evolution in man.

And consistency demands the same conclusion. If birds are 'evolving' — developing, advancing in different ways and at different rates— then some birds are further along, are more evolved than other birds; some, more fit than their (un)equals.

Then, what's good for the goose … . What's true in one species is true in another. Bird, human—the same. If we are all evolving at different rates, then we're all on different levels, some, further along in development: individuals, races, the truth is the same. Some humans, as some birds, are more evolved than others.

Would you mind repeating that, Adolph?

Not nice. Not true. But a fact you're forced to conclude if you're honest in your use of your terms.

Adaptation, and variation—beak shape and size, length of tail feathers—these traits do not represent kinds of animals evolving into other kinds of animals (the finches remain finches). These are modifications springing from the rich genetic diversity already within the kind; modifications which come to the fore as species adapt, both to their environment, and to the changes in their environment.

Variations from *within* the kind: not one kind jumping to another.

Reflection 15 – Vestigial Structures

As far as newborn Mary's aware,
every body member is useless:
no part with purpose – all
organs, all structures,
vestigial.

But Mary grows in self-awareness,
gains some education:
a Masters and a PhD,
Mary gets an earned degree:
Biochemical Concatenations
of Human Physiology.

She says she'll keep
her thyroid now, her
tonsils, and her thymus,
her appendix, and her
pineal, her
adenoids, and her sinus.
And since she never
wants to lose
her pelvic diaphragm,
she's gonna keep her coccyx, too,
"These parts are wholly
who I am."

Now armed with facts,

she struggles; sad —

takes all she's got to be Mary now.

We're certainly a proud lot: *If I cannot explain what a body part does, its function is quite unnecessary.*

But as our knowledge of 'vestigial' structures has increased, so too has our appreciation of all that they do. In the last hundred years, the number of 'useless' organs, structures, and appendages — evolutionary leftovers, we're told — has decreased from 180 to 90 to 10 to ....

Seems all our parts have been serving a purpose, whether we knew it or not.

And wouldn't you know it? We're even running low on our 'junk' DNA.

Reflection 16 – Irreducible Complexity

Deer season here in my field –
wearing Hunter's orange, but
bearing no gun. Yes –
once,
survival: existential
constraints. Take the
buck – the
meat, the
fat, the
hide; take
every last bone for broth.
All needed then, but,
now?
Let the deer be; give me
glimpses –
honeyed streams of kinetical grace.
If taken only for its rack,
if just to hang its head – hunters,
you should do the same.

These entrails at my feet –
field dressed; they
left his liver.

You could eat liver once: not
now –

this Good Earth racked
with poison.

I left my liver in San Francisco
Love the alliteration, but you'll never hear it sung.
Too bad about the liver; it rarely gets its due.
The human liver performs over 500 vital functions. It produces bile (which breaks down fat for better absorption), converts ammonia (that will kill you) into urea, controls blood clotting, keeps your eyes from turning yellow, produces proteins which transport fats, regulates blood levels of amino acids needed in protein production, purifies the blood, stores and releases glucose, processes hemoglobin, stores iron, stores glycogen, carbohydrates, and vitamins A, D, E, K, B12 – and on and on we go.

The liver's the only organ that can regenerate itself. Cut away up to 75% of it, it'll all come back. But here's the kicker – a fantastic fact about the lowly liver: There's no loss of its many functions while it's in the process of regrowth. (Take THAT, you simple heart, you.)

Over 500 functions. Vital functions. (Oh, the meaning of words.) That's vital, as in *vitalis, vit, vita* – life. As in, if a function's not performed, you sicken, you weaken, and then you die.

250 working functions won't do it.

350? Sorry. Not enough. All operations need to be working, and all of them working at once. You can't – as gradualistic evolution would have us believe – start with a few functions firing while you wait eons for hundreds more to come online.

The origin of cellular transport, blood coagulation, antibody diversity: Irreducible complexity says, for this to work, it *all* has to work—together; at once; and, now.

Not so! I'm a product of piecemeal developments spread over three million years!

Okay. Then tell me. What piece is in development now?

Reflection 17 – Eyes and Ears

The barn's gone:

that spring-toothed
harrow
propped against a wall, chewed
rails in the horse stalls,
lovers' initials
carved in every post;
the owl in white-face
perched on her beam –
no need of eyes, her hearing
so sharp,
mice,
hidden – quarter-mile distant, and
tucked under straw –
plucked from their beds
by one on wings so silent,

she cannot hear herself
fly.

The poor man had trouble with his eyes—at least one would think so, from what he told his optometrist:

You know, Doc, to suppose that the eye with all its inimitable contrivances for adjusting the focus to different distances, for admitting different amounts of light, and for the correction of spherical and chromatic aberration, could have been formed by natural selection, seems, I confess, absurd in the highest degree.

"I know it's absurd", said Charles, "but it's the only explanation I've got. I'll trust man's reason to see him through."

Human eyesight. A big blur? Then drop down a level or two. Have a look at how the lesser-evolved see:

* The eye of the squid, with its lens, its pupil, its optic nerve — nearly eyedentical to ours. (No development here from a light-sensitive spot on the skin.)

* Go lower still, to the trilobite — the earliest group of arthropods. Trilobites are in a class of their own: extinct mud-dwellers, true bottom-feeders. They range from the microscopic, to just over two feet in length. They're found on every continent, and in the lowest of the fossil layers.

Yet, look into the trilobites' eyes. Varied in form, but always spectacular. Eyes with 360-vision. There are eyes on turrets. There are stalked eyes — at the end of tubes, and all of them set at slightly different angles. There are eyes with separately inlaid lenses of clear calcite, exceptionally pure — no distortion. Some eyes are comprised of 10,000+ lenses, packed in patterns of hexagons.

Amazingly complex eyes for simple bottom-feeders, those whose station in life was below the first rung of the ladder.

Reflection 18 – Homology

I met my neighbour returning from the trail the other day. Head down, mumbling, she was in some distress.

"Pardon me, ma'am, but your horse is missing a leg."

"I know. I know. I didn't see – poor Rufus – I didn't notice at first, but I think ... it might have dropped off in the birch grove. You heading up? Keep an eye out for it, would you? I've got to get him home. I don't know what I'm going to ... oh, watch for it, would you?"

"Will do."

I go to the grove. Look around. I don't see a leg, but I do find a bat wing. I lay it out on my palm, and come back down the hill.

I knock on her door. "Here you go. I found this. It's not the same size as the one you lost, but I'm told bat wings and horse legs are homologous structures. So ... cheer up, Rufus," I yell across the paddock, "I know it's not a true leg, but it does have a similar design. Any vet worth his salt lick should be able to attach it. Your limp should disappear in ... oh, say, 35,000,000 years."

Homology is a term which states that animals with similar structures must be closely related—must have come from a common ancestor. Since a bat wing and a horse leg have a number of similar bone placements, they had to have come from a forerunner. (Poor Rufus. He's only a three-runner now.)

Similar structure—common ancestor: from the same branch on the tree of life. That's evolution's read on homology.

Okay, but ... what of the myriad common structures found in the animal kingdom?

As noted above in the squid, the octopus, too, has an eye very similar to the human eye. Both have lenses which project images onto a retina, sending impulses to the brain. Similar eyes—it's spooky how similar they are. Surely, an example of homology?

Or, the pig's heart? Very close to humankind's. (Some of us have their valves.)

Or, the red blood cell-count of the frog? A good match-up with humans here, too.

These must be cases of true homology — creatures with features so congruous.

Nope. Not so fast. One would think so, but, no.

Why not?

Here's what we're told. (You'd better take a breath.)

We've put the octopus on a tree-of-life branch which is ages removed from man because the common ancestor of cephalopods and vertebrates had only a light-sensitive skin patch. (We never could have gotten such a complex eye from anywhere on this branch of the tree.) So, even though the structures (eyes, hearts, red blood cell-count) appear to be very similar to man's, because the 'families' (man/octopus; man/pig; man/frog) are so obviously different, we must refer to these similarities by another term. We call this convergent evolution — similar structures that have evolved independently in animals that are on different twigs of the tree.

So, what we're being told is …

Similar structures are examples of *homology* when animals are located on the same branch of the tree (a placement given on the basis of their possessing a certain similarity in structure from their descent from a common ancestor) but, these same similar structures are examples of *convergent evolution* when animals are located on different branches of the tree (a placement given on the basis of their dissimilarity in structure from not having a common ancestor in the first place).

Okay, then. I see (no matter where my eyes came from).

But I'm not to question such circular reasoning?

Reflection 19 - Fossils

Oct. 15

Heavy rain.

Feet plough windrows of leaves.

Oct. 18

Third day of streaming sun.

One, red, maple

leaf —

margin, sharp,

full thirteen tips,

fit trophy for plaster cast,

or fossil

fixed in stone.

That dead crow:

don't disturb it.

A wet, fall day. A quiet walk, feet cushioned on mats of sodden leaves. Won't be much left of these leaves come summer: winter's freeze, spring's thaw — will break them down, will return them to the earth.

That crow, too, eventually — whatever's not eaten by maggots. Or coyotes. Or vultures. (Yes. We have them now: buzzards on the Nashwaak. A strange sight, those gawky-necked scavengers circling the heavens. Welcome to the Arizona buttes.)

Nothing dies and 'just becomes' a fossil. It takes some very special circumstances, some special conditions, for a leaf, for a crow — for a monster T. rex — to fossilize.

You need a proper mix to keep them from wasting away: water, sediment, minerals, mud. No oxygen. Though some victims are encased in resin, others go through petrification—a crystallization of minerals replacing original material.

Every fossil has come through a narrow window. Burial must be quick. Left to the elements, the dead soon disappear. Everything has to be going your way before you can turn to stone.

But fossils aren't exclusively stone. Increasing numbers of them are found to contain soft tissue: blood vessels, skin, red blood cells, and collagen. Scientists are finding remnants of original biochemistry over the entire spectrum: in insects, worms, birds; in amphibians and reptiles (seems even T. rex has a soft spot or two).

Soft tissue? From fossils found in rock said to be up to 500,000,000 years old?

I'm not to question how this can be?

That poor crow—the target of microbes, maggots, foragers of feather and fur. He can't remain exposed if he's hoping for mummification. Though I'd love to see his skin and bones go the way of Egypt, I know that all soft tissue must go the way of grease.

Reflection 20 – Boneyards

The dogs are not with me today.
Glad of it – this
rabbit, laid
dead on the path, half
hidden in grass,
skidded me slick into clover.
Can't tell why he died,
poor fella. Alone
on the trail. Alone
in death. Alone
at the end. Alone.
You foxes, you
coyotes, you dark
vultures, all,
leave the poor thing
as he is

alone.

It's rare to find a fossil by itself.

Think 'corpses'. While there's a slight possibility of finding one buried in your rose garden, you'd improve your chances immeasurably by shifting your dig to a graveyard.

It's the same with fossils. Sure, sometimes there's one here, one there, but the bulk of fossil finds are just that—in bulk. They're found in fossil graveyards; boneyards, we call them—mass burial plots dispersed throughout the world.

And, my, there are some big ones:

- The Hilda Mega-Bonebed, Alberta
- Burgess Shale Deposit, British Columbia
- Stonehammer Geopark – right here in New Brunswick
- Joggins Fossil Cliffs – Nova Scotia

You can see that Canada has its share of buried bones. But the finds don't stop at our borders. Important international sites include:

- Green River, Wyoming
- Hell Creek, Montana
- Jurassic Coast, Dorset, England
- Seymour Island, Antarctica
- Ediacara Hills, Australia
- Zhucheng, China

The size of these yards runs from a few square kilometers, to what we find in the Karoo Basin Fossil Field, South Africa. The Karoo Field stretches over 200,000 square kilometers (just slightly larger than your old church graveyard).

And the numbers contained in a singular bed? From hundreds of dinosaurs, to thousands of fish, to multiple billions of nautiloids.

Scientists are finding all sorts of interesting (and puzzling) things:

- Clams fossilized with their shells still closed. (But clam shells open in an 'ordinary' death.)
- Fish fossilized in the process of giving birth.
- Fossils of animals with undigested food in their belly.
- Animals fossilized with their prey half-swallowed.

Impossible 'death scenarios' under slowly accumulating sediment.

Reflection 21 – Catastrophism

There's a river at the foot of this hill. It once carried salmon; it once floated logs to the mills.

There was a day when 600 sawmills buzzed the banks of New Brunswick's waterways – a day when the white pine reigned majestic: heads through the clouds, feet cut from under them, masts for the brigs once built in Saint John.

You needed the right weather, and at just the right times, to get the winter's cut to the mill: ample snow for the sleds (though not to a depth ceding six-foot stumps), and a steady cold for a base. A January thaw was a thing to be feared – could kill both man and beast. Yarding in a slush-and-muck mix could explode the heart of a stubborn horse. And the flash-freeze that followed: a harvest of busted bones.

But both snow and rain were needed for the run. A deep-enough snow-pack locked in the woods, waiting for a warm-enough rain. Just enough rain to melt just enough snow for just enough water come freshet. A surfeit of water, from rising too quickly, would strew the floated timbers far and wide. But that same flow dropping too rapidly, would leave the logs stranded – first usage of 'left high and dry'.

But low water, too, had its perils. Logs clogging the riverbends increased the chances of jams. Here the flow stopped, but what was behind kept coming, loggers now in the shadow of a burgeoning wall of wood.

Such a scene set the call of the dance: Stout men stepping in a butterfly ballet – flitting log to log, armed with peaveys – straining to drive their hooks deep into the jam. Gridlock too solid? Sound the call to Partner Dynamite waiting in the wings.

Whether on the water, with its variations on the themes of direction and speed, or on the skidways with the logs stacked on cribwork, lumbering was a dangerous game. One slip would do it: broken bones, mutilations – a man crippled for life; the ever-present spectre of death – either by drowning or 'nail-kegging' – the logs turning end-over-end, crushing any man beneath.

MacBean Brook: named after John Angus MacBean. A brave brute of a man, respected by all, but one whose fame came at too high a price — having pinned his name to a stream.

It's not only that fossil death-pits are found throughout the world, it's the conditions in which these fossils are found that puzzle the paleontologists.

Remember our 'perfect specimen' red, maple leaf? Most fossils aren't so lucky. Yes, thankfully, some are in near-perfect condition, but most are found cracked, broken, or crushed: crash scenes resemblant of thousand-car pileups. Not what the scientists were expecting. How could gentle layerings of sediment cause such deformities?

'Perfectly preserved' is not the general rule. Most tell a different tale. Bonebeds reveal fossils entombed in mass-burials, and in just-plain-weird conditions: frames stacked, piled on, piled up (in some places, two hundred and fifty feet deep), corralled into valleys and caverns. Whale carcasses heaped in high deserts. Dinosaur leg bones — not merely broken, but split down their length. Two-ton creatures in piggy-back. (Quite a balancing act, this — propped in tiers, waiting to be covered in silt.)

This is what we see — what the science shows. Sudden. Destructive. Entombments.

Swept away to be buried alive.

Reflection 22 – Uniformitarianism

Thomas Murray – that old recluse – accompanied every birth on the river over a course of seventy years.

No matter how distant the house, he would arrive just after the baby did – gift in hand, every newborn presented with a just-picked bouquet. Even out-of-season comers posed no problem for Tom, for he'd unearthed the secret of drying flowers.

And we're not talking crunched clover here – simple three-lobers pressed into pages, leaves between leaves, as it were: these, real flowers; flower flowers; the colours, the form, Tom's was a method of full preservation – each floral attribute, not just preserved, but enhanced. Flowers as full as a carnation, flowers as sensitive as a bachelor (now there's a rare thing) … button – as a bachelor button – their petals staying vibrant for years; dead, yes, certainly, but in colours brighter now than the day they were picked. Tom's trademark throughout the valley: fresh to those born when flowers graced the meadows, dried when heads were in seed.

But then there was that episode with Margaret MacBean.

Margaret was heftily overdue – three months at least, by her husband's calculations (only three days by Margaret's). No matter the count, her final big day finally came.

But Margie'd already been booked! A full year before – on this very day – she'd been scheduled to be delivering, yes, but a speech – a baby, the furthest thing from her mind.

And so now that the day rolled around (Margaret doing the same), she waddled next door to the IODE Hall – keynote speaker at the gathering of the Daughters of the Empire. There, straddled betwixt pillars colonial (what else?), she was waxing poetic, when, dropping down on page eighty-eight, she continued the descent to the floor, engulfing the hall in screams.

It was a good thing Nellie, the village midwife, was there.

And that Nellie's granddad, David, had survived the Battle of Balaclava.

And that he took Nellie in when both her parents drowned.

And that despite his being wounded in the ... Crimea, David gave Nellie a loving, supportive, home. He became her inspiration, though she struggled at times with her grandad's habit of framing nuggets of wisdom in militaristic jargon based on commands given in the charge of the Light Brigade.

Still, it was to serve her well now. "Hot water and khaki," barked Nellie. "You will have this child, Margie, but you're gonna have to push. There. Nearly there. Keep pushin', dear Margaret — half a leg, half a leg, half a leg onward."

As promised, Margie made it. So too, the child — for a minute. Poor thing. Held only a dozen breaths. Some said it was a girl.

And Tom made it, too. No one knew how he knew to head out that day, to walk the four miles with a readied bouquet, arriving one minute postpartum.

But he carried no colour this time. A simple gift — a twig tip of an oak; a rosette of dark, dead leaves encircling a cluster of acorns.

The midwife wrapped the child. Cradled her into her mother's arms.

"Margie," Tom whimpered, "I've been sent with a message." He folded an acorn into her palm. "Life won't always be like it is right now."

Uniformitarianism. Quite a mouthful. What does it mean?

Uniformitarianism is a theory which states that Earth processes which are active now, are the same processes which operated in the past, their exertions through the ages, continuous and uniform — hence, uniformitarianism.

What you see now is what has always been.

The present is the key to the past.

This supposition is held across several disciplines: paleontology, anthropology, evolution, and geology, all hold to some aspect of uniformitarianism.

Their tenet: Since physical processes must operate in a consistent manner—at essentially the same rate and intensity as they do now—by extrapolating the knowns, we gain a view into the past.

Sounds simple enough. But the theory has significant problems.

If processes have always worked as they do now, how can it be that fossils of marine invertebrates are found on the world's highest mountains? There are no actions or energies performing such feats today. Yet, the science is clear; the fossils are there. Something had to have been different back then.

A global ocean five miles deep? Continents in crunch, driving ribs of rock through the clouds? It has to be one or the other. Maybe both. But to claim the present holds the key to understanding how things occurred in the past? Not so fast. Whatever singular dynamism it took to raise mountains is not being witnessed today. How is this a uniformity that runs through the ages?

And what of the tropical-life fossils found at the poles? Conditions were much different once; not what we see at present. There are no fields of fossil deposits being formed in lakebeds today. They obviously were in the past. Formed then, but not now. Where is the continuity?

If the present rate of decay of the Earth's magnetic field is the rate that has always been, that protective covering would have dissipated long ago. The fact that it still surrounds the Earth speaks to the mathematical necessity of a shift in the rate of change; it couldn't always have been what it is now.

Or, this: Try pointing to the magnetic north pole. Make it quick. It just moved. Again. Always. In flux. Not only as to its location, but in its acceleration as well—in its rate of change in velocity. The once 15 km/year shift has increased to 55. How can this be said to be a constant rate of change?

The challenges raised by these scientific facts cannot be 'corrected' by claiming every theory makes allowance for slight modifications. Nothing slight about it. Uniformitarianism has serious problems.

One of the biggest? Many fossil beds show every major animal kind appearing together in the same find. A real dilemma, this one. So how is this nail-in-the-coffin-to-gradual-development-of-kinds explained (away)?

We are told that all of these kinds only appear to have appeared at once. (I'll wait why you go back and read that sentence again.) You read it right: They only appear to have appeared at once.

Here's the explanation:

These animals still evolved in their kind-to-kind sequence, but there was a speeding up of the evolutionary process, a sudden variation in the rate of change — like a snap-snap-snap of the fingers; everything sped through the processes resulting in the links dissolving completely — then slowed again to the normal evolutionary rate of change. This is how they all ended up in the same place at the same time.

Oh. Okay.

So ... more of a selective uniformitarianism, then? Sometimes held? Sometimes not?

And I'm to take this as serious science?

Reflection 23 – Cambrian Explosion

June 11, 1819.

Sixty-three days from Wales to Saint John.

Cleared quarantine,

held service on the Fundy shore.

One hundred eighty gaunt worshippers.

No chapel.

No pulpit. And no plate passed –

nothing to give

but thanks.

Pressed on

up the river, afloat

on the promises –

third tier Land Company plots,

turn-key cabins,

a mere fifteen miles from town.

Griffith, Evans, Lewis, Jones –

collectively, Cardigan Settlement,

though teetering on the Edge of Precarious

would better define the day:

no staples for winter,

October snow,

willow-bark soup,

no skills in milking

an iron breast of a frozen earth –

though death, a breeze,
the Company reneging on windows and doors.

Sympathetic hearts:
the Fredericton Emigrant Society
carried families to town,
set them in sheds and barns.
Gaelic and English —
sparse interchanges, but
common themes:
shelter, food,
a smile from a child.

More sailings from Cardiff —
shoulders to the wheel.
Still, ten years
to bring Cardigan to life.

Life always roots.
Life always blossoms.

From the back of the Lewis' farm,
he cut a trail to follow the watercourse —
Tay Creek to the Nashwaak
River;
seven miles through.
No distance at all
for a young man in love —

Robert Lewis wooing Nora Doyle.

Wales, or Cambria: Cymru in the ancient tongue.

Coal, castles, choirs, and chanties. But anthems weren't the only sounds heard from the hills of Wales.

In the 1830s, there thundered one mighty explosion. Not from a build-up of gas in the mines, but it did come out of the rocks—from a find in the fossil beds.

'Explosion?' Yes. (Though the word's metaphoric, it sounded from stone sedimentary.) The Cambrian Explosion—a discovery that continues to shake the paleological world: the abrupt appearance of every major animal group in the lowest level of fossil-bearing rock.

Bears repeating, but let us come at it from another direction.

Bring your shovel. We're doing some digging. But this time we're starting at the bottom of what will become the hole: not digging down—digging up. From way down there, we're heading to the surface.

Okay. Start shovelling. Up. What do you find? Rock. Rock. No fossils. Rock. Keep digging; we're heading to the surface. Rock. No fossils. No fossils, then BOOM. You found them. Fossils everywhere. You've come into a layer of fossils. And (here's the 'explosion' part), you've discovered not only fossils, but fossils of every major animal group. All at one level. All at the same time. No transitional forms. No evidence of common ancestry. All groups unique, clearly identifiable, with full, complete, and (at one time) functioning, structures.

There's your explosion. Talk about your big bang! Hypotheses, theories, assumptions, interpretations—all heading for the hills at the speed of light, hoping for a black hole to hide in.

And, while it's called the Cambrian Explosion because Wales was the place of first-discovery, these fossil-bearing levels are not confined to Wales. 'Cambrian discoveries' have been found the world over: Montana, Wyoming, western Canada, eastern Russia, and southern China.

To quote Sir Charles, these finds present a 'serious difficulty' to the theory of evolution.

How could all life have developed from a single line over millions of years, when trilobites, worms, coral, clams, jellyfish, and fish (yes, even vertebrates) are all found in the same time frame, and at the same location?

And with no ancestral forms amongst them? Or below them?

In-the-field scientific observation uprooting the evolutionary tree-of-life hypothesis.

And I'm not to ask any questions?

Reflection 24 – Transitional Forms

No one on the Nashwaak knew Ciara's age. Martin Irwin claimed she had always been old – born into this world wrapped in wrinkles. Not the soft undulations all babies come clothed in, Ciara's skin ran in ploughed ridges, as furrowed as windrows of straw. Ciara Smith had always been old.

So, eighty-five? Ninety-five? Was anybody's guess, and everybody did. Peter Murphy had her ciphered as 107. Said he'd seen the church records. Could've proven it, too, if some thief hadn't stolen the journal.

But Ciara was more than her years alone. She flowed forth a composite – tributaries streamed from her ancestral pool; a thousand lives poured into one. For Ciara was a traveller – of walking people stock: mystery in overcoat, spectre in shadow, as steam clings and rises from thick, wet wool, residuals of her presence hung pungent: pine, on forest trails under new moon; cedar, in misted damps; soil, on a grave just filled. And what of her angry spurts in old-country lingo? All the proof needed she spoke with the dead.

She kept the largest herd of cats in the county – all of them, black; only half of them, dead. She could caw the crows down from the clouds. They'd hop in circles round her feet, servants at the wing, fulfilling their Majesty's wishes.

Ciara knew secret things. Born with eyes sewn shut in mucous webs – took ten days for the gunk to dry, to crack as fondant, to fall off in scales. Was this the source of her second sight? She predicted Peter's daughter would fall through the ice; spoke of sparks in the wind that carried music; warned of a coming great war where millions of mothers would lose their sons.

Few got invited to tea. Those who did, spoke of stuff – stuff, and more stuff. Told of ploughed paths through the clutter. Of upholstery sliced and gutted. Of walls pocked with holes – sprays of ack-ack in lath and plaster.

"I'm searching for a goatskin," she'd tell them, " – I'll get to this mess when I find it. My family's guarded that scroll for thirteen generations. I'll not be the one who lets slip the charge. I was reading *it to Grandma when she ... shifted. The potions, the wisdom – I cannot die without it. Have to keep digging. I'll clean this up after I find that skin."*

Alas: May 29, 1919 – furrows to furrows, as Ciara had insisted, buried on the day she died. And on the total eclipse of the sun – that one black ribbon weaving its way through Peru, Bolivia, Brazil; on to Liberia, through the Congo; leaving Zanzibar, slipping into the Indian Ocean – knitting together every village where Ciara had lived a lifetime.

Only a handful gathered on the Nashwaak that day. A final hymn. As they lowered her into the earth, a lone crow gurgled from an overhanging oak.

Neighbour Margaret left the grave, went to feed Ciara's cats – found the house preened to a feather.

It's a simple thought, really.

If all life has sprung from a single cell—has developed from that one common spark—then logic, science, (and even common sense), would demand an immense number of transitional forms—links connecting the group before to the group that follows. This is the foundation of gradualism: the slow, continuous evolution of species over long periods of time. It would absolutely have to be this way.

Mr. Darwin said the number of transitional forms must be truly enormous—every stratum of rock should be full of intermediate links.

But there's a problem.

250,000,000 catalogued fossils later, no 'immense number' of transitional forms has been found. No 'small number', either. Not one true, clear, obviously transitional form has ever been dug out of the earth. Not one.

But they must be there; they have to be there. The evolutionary hypothesis demands it.

But they're not.

So ... what to do?

No problem. Enter the *hopeful monster* theory.

It goes like this: A crocodile lays an egg; it hatches and out pops a bird. Now, a baby bird might sound cute to you, but to a croc, a bird is a monster. Yet I'm told this is how humankind could have evolved. Since no transitional forms have been found (*absolutely necessary* transitional forms), then whenever gradualism needs to fill in a gap—as in jumping from kind to kind—a monster suddenly appears.

The *hopeful monster* theory, though losing ground, is still a contending theory as to how mankind first appeared.

Okay. Then half-points.

I'll give you the croc.

Reflection 25 – Geology

I made a few phone calls; I knocked on some doors. Someone in St. George had to know where it was.

The lady washing her truck said, "Yes. Right there." She pointed across the road. "That lot runs to the river. There used to be a building – a barn, a shed – whatever the place is called where they finish the stone. Stood there for years. There's still a pile of granite on the river bank. Some good-sized pieces, too. Just step over the chain; my son owns the property now. You should find something there."

I thanked her and crossed the road. Went down to the river and picked through the pile. Found tons of granite at the water's edge. I took two dozen pieces – different colours – red, orange, black – but each piece I chose had one thing in common: a polished face.

I wanted proof each piece had been touched – held in a stone master's hand a hundred years ago.

I love rocks. I collect them from everywhere. I've got Bay of Fundy rocks in my garden. Got Newfoundland rocks there, too – over sixty pounds of pickings, and a hole in the bucket to prove it.

I build with rocks. I landscape with rocks. I play with their lines, their colours, their shapes; I lay them in patterns and pathways. I will stand in the midst of rockwork and say, "Wow. Look at you – how you anchor all of this! Why would anyone bother with plants?" (Okay. We ALL need plants, but ... you know ...)

Drive hundreds of miles to admire a drystone wall? Yeah. I've done it.

I love rocks.

But there's a rock in my garden that shouldn't be there. Yes, it's mine. And no, I didn't steal it – in fact, I paid for it twice. Gave twenty bucks to Tom (the guy who owns the pit), plus, I crushed a new dolly while one-manning it into my truck. But I had to have that rock; I had to take it home.

It occupies a prominent place in my yard. Surrounded by a ring of lowly rough-n-readies, my rock reigns from its rightful throne. Though not much for colour—grey straight through—its skin's as sleek as poured silk.

It's impossible to walk by and not touch it. But you can tell it doesn't belong.

It's not the only rock that shouldn't be where it is.

How can rocks which contain complex life forms be in the lower levels of strata? Shouldn't they be near to the surface? If higher life developed through the process of gradualism—evolving from simple to complex—and, if fossilization comes from the slow accumulation of silt, shouldn't only simple forms be found in the earliest layers? And for these to be found with species which should not appear for another 10,000,000 years? Something's out of place.

600,000,000-year-old limestone on top of 100,000,000-year-old shales? We're told the older rock is an overthrust—was pushed up and over the younger. Okay. Makes sense. But shoved for a distance of 50 miles with no significant evidence of rupturing, or of erosion between the layers?

We see sedimentary layers—running in wavy horizontal lines, bending into angles and curves. How could such formations be possible? Huge volumes of silt would've had to have settled out rapidly, undergone upheavals of sufficient strength to curve the still-pliable layers (else the bends would be cracks), and only then, hardened to stone.

How can two adjoining layers of different compositions—sandstone and limestone (both having been bent when soft, and bent at identical angles)—be said to have an age difference of 300,000,000 years?

And what of the polystrate fossils these layers contain? Polystrates are fossils running through more than one layer of rock—an individual fossil, the very same fossil, found trapped in more than one layer.

To see a fossilized dolphin—bottle-nose down, tail to the sky—spanning three different strata, and to be told each layer took a million years to form, strains credulity. Standing on its head, exposed to scavengers and the elements, left untouched, not rotting away over millions of years?

And what of the *geologic column*, that cross-sectioned, top-to-bottom cut-through of various rock layers and the fossils they contain? (You've seen pictures of the exposed faces.) It's said to have formed by sediments laid down over eons. Fossils and Time. Slow. Settled. Everything in its place.

Okay, but everything isn't. While it may be a helpful tool, the so-called global column isn't global at all—a full two-thirds of the Earth's surface showing less than half of the ten geological periods. The column is merely a construct—a composite of layers based on extrapolations; a diagram of incorporated strata, taken from the world over—a puzzle of pieces and thought.

Hear the circular reasoning behind it:

We know these fossils are 300,000,000 years old because they are found in rock which we know to be 300,000,000 years old because of the 300,000,000-year-old fossils they contain.

Doesn't sound like good science to me.

No matter.

I'm keeping my rock in my garden.

Reflection 26 – Carbon Dating

A thousand rocks inhabit this pile hunkered at the edge of my field. Just another phase in their journey. Dormant. Disturbed. Exhumed – supplanted by beets and potatoes. From beneath the earth, to beneath the sun – skin discolouring, thickening: from a damp, smooth, grey to a parched green-black, etched in scrawls of lichen.

But what are these pieces of ... are these chunks of coal? Could these be residuals from the 1890s? From where the Little Tay meets the Tay River; where Jed Segee, the village blacksmith, once mined fuel for his forge?

Jed's seam was a finger of a much larger vein. Dig out your map: Durham to Minto – a straight run through the woods, twenty-five miles for a crow. A hinterland bound by the Cains and Salmon rivers – twelve hundred square miles of nothing but trees, or scars where trees once stood. It's still back-country. You'll run into no towns, but you will trip over seams of coal.

Minto: the first coal mine in North America. Heated homes clean to Boston – rather, clear to Boston; there wasn't much clean about it. But after being dug for 375 years, her coal was labelled too dirty: laced with mercury and sulphur. They closed the mines. Tore down the power station. Pretty much shuttered the town.

But jobs aren't the only things not adding up.

The recipe for coal: Let dead plants go to peat, add heat and pressure, then sit back and wait for ... a period of time.

A mixture of hydrogen, oxygen, sulphur, and with a fixed carbon content as high as 85%, coal is clouds of smoke, blankets of soot, black lung pulling your breath through a straw; worn-finger smudges on all you touch.

But coal is warmth. And coal is steam: energy for industry, heat for homes. And, carbon's what you want when you're dating – estimating the age of organic materials.

Carbon dating. How does it work?

A radioactive isotope of carbon — carbon-14 — forms in the atmosphere from cosmic rays hitting on nitrogen. Since plants need carbon-dioxide (CO2), when they take in carbon-12, amounts of C-14 are 'ingested' as well.

But when it comes to dating, carbon-14 needs to get a life — it's only got a half-life now. Half-life: The time it takes for 50% of a quantity of radioactive material within a substance to break down, to decay. Radiocarbon (carbon-14) has a half-life of 5700 years. This means that when half drops to half, drops to half, drops to half, detectable amounts of radiation disappear in less than 100,000 years.

This measurement of decay can be helpful. If you know the initial amount of carbon-14 in your sample, and if you know the sample hasn't been contaminated, and if you know that the rate of decay has been constant (yeah — a lot of 'ifs'), you can estimate the age of whatever it is you're holding in your hand.

But all is not as it seams. (Leave it be, editor. We're mining coal, remember?) Carbon dating can only be used in organic material — wood, bone, shells; it won't work on rock itself. And it's only detectable within its half-life range: 100,000 years.

So what does the science show us? We find measurable radiocarbon in coal, in diamonds, in dinosaur bones — fossils said to be 120,000,000 years old, well beyond the range of detection. Something's amiss. Either it's in the methodology (No. The science sounds good), in the instruments used in detection (Could be. Check your calibrations), or, it's the a priori assumption of age (Ah-ha. A date worth a second look).

Other methods are used for dating inorganic substances. Rates of decay are measured as uranium drops to lead (U-Pb), as potassium drops to argon (K-Ar); rubidium to strontium (Rb-Sr), samarium to neodymium (Sm-Nd).

Again, any true results depend on a known rate of decay, a known initial amount of radioactive material, a certainty of non-contamination, and a known uniformity (there's that word again) in the rate of decay back through time. Line up these variables, and you should get the age of your sample.

Instead, we get some serious problems:

* The Kilauea volcano erupted in 1959. Lava cooled to basalt, an extrusive igneous rock. Radiogenic readings determined their age to be north of 8,000,000 years.

* Mount St. Helens blew her top in 1980. Yet the dating methods tell us that the dacite rocks formed from that explosion are 350,000 years old.

* The Mt. Ngauruhoe volcano, New Zealand: identical rock samples were tested by all four of the above methods. The declared ages ranged from 270,000 to 3,900,000,000 years.

So, still looking to make a selection from the dating pool?

Take your pick — the only tool you'll need.

Someone else is doing the shovelling.

Reflection 27 – Scientific Method

If you chew these veggies ... very ... slowly ...
you'll have to spit them out

Some times I sit,
sometimes I sit and
think: I'm
forced to think, for
even a cabbage comes with a head; a
potato – multiple eyes.
I think, therefore I yam.

But when I do think – when I
chew some tidbit
fit for food-for-thought,
the pulp proves most unsavory.
Celery stalks
my conscience.
And broccoli spears
my heart.

Where did it come from,
that pre-existent seed
– matter, energy, space, and time – that
prototypical speck;
one finite grain of all that would be;

*ten million billion trillion
tonnes
pressed to infinite density?*

*No one there to see it.
'Non-observable'.
Only happened once.
'Non-repeatable'.
No dissemination for peer-review.
'Non-verifiable'.
And yet I'm fed answers said found
on the road of ... scientific ...
discovery?*

*H'm.
Repudiating the steps of methodical science?
Forcing me to swallow
your propounded explications ...
on faith?
I'm a product of time?
A victim of chance?
A creature a-slog in the cosmic pulse?
(Radiation or peas – won't matter.)
A dis-gorge of sunshine and
hydrogen soup?
No reason for me
to sing Joy to the World?*

Then, please wait for me, Ms. Mitchell.
For I wish I, too,

had a river to skate away on.

Going to Kansas City? A hundred roads will get you there — come at it from any point on the compass.

The same can't be said if your destination is empirical science. While it's a wonderful place to get to, it's not always easy to reach. Has a much more limited access. And you must follow directions precisely. Obey all the signs. Stay within the lines. Keep a record of your travels. Text a detailed map to your friends so they can get there, too.

In a very real sense, empirical science is more about the journey than it is the destination. And what's interesting about it, is this: While you're driving down the road to get there, you'll find the traffic flowing in both directions. The way to get to where you're going: inductive and deductive reasoning.

Pull over for a second while we learn the difference.

* Inductive reasoning:

You move from specific observations to form a general conclusion. But you have to be careful here. Your conclusion may be probable, but won't necessarily be true.

For example: I observed it was sunny on Monday. I observed it was sunny on Tuesday. I observed it was sunny on Wednesday. So, on the basis of all my observations, I conclude that tomorrow, Thursday, will be sunny as well.

This may, or may not, prove to be true. Based on observations, the chances are pretty good, but ... well, you know how meteorologists are.

Or, I observe that Peter has big ears. Peter is from Toronto. I conclude that all Torontonians have big ears.

Yes, inductive reasoning might be helpful in planning picnics, or sizing a pair of earmuffs should you need a gift for a blind date who hails from Kensington Market, but you have to carefully weigh your conclusions. You can form hypotheses and theories — even state some probabilities — but they may not be necessarily so.

* Deductive reasoning:

You come at it from the opposite direction. (Two-way traffic, remember?)

In deductive reasoning, your conclusion has to be true if the premises are true. Your reasoning follows a logical path. You get to use the word 'Therefore' in deductive reasoning.

Peter lives in Toronto. Toronto is in Ontario. Therefore, Peter lives in Ontario. Deductive reasoning is based on observable truth. (I know) Peter lives in Toronto. (I know) Toronto is in Ontario. Therefore, (I know) Peter must live in Ontario.

Both lines of reasoning are helpful; both have their place. But did you notice that both involve *observation?* — not something you can do with the past. There's no way to observe what the Earth's processes were 4.7 billion years back.

You can speak of 'probabilities', but you have to be cautious in calling it 'fact'.

Because sometimes in rains on Thursdays.

And, "Hey, Torontonians! Lend me an ear. Don't worry about not being able to hear. I'd say from the size of things, one ear will do you just fine."

Reflection 28 – Unfalsifiability

1893 – a hard year in the valley. Poor harvests all round. A late October blizzard that never let go. Twelve feet down by February – the same month the Tierneys and Bubars had their homes swallowed by fire.

While the village had always known troubles, this batch, a heavier mix. Brought its own blend of sadness: settlement in sackcloth.

So, the announcement: To lift our community spirit, an Irish stew night will be held at the church. Though no one has much, bring what you can. Those with little shall share from surplus; those with nothing shall know no lack.

Saturday, February 25th. The day came in cloudy. Heavy snow by mid-morning – thick and thicker by noon. When it came time to harness the sleighs, to load up the cooking, you couldn't see your pans in front of your traces. Still, the meal – tendered for the social and spiritual benefit of all – would go ahead as planned.

Nine families with forty-two kids set out (forty-two in total, not each), and nine families wished they'd never left home. Had nothing to do with the weather, everything to do with a storm – a slow-moving disturbance named Cyclone Seamus O'Neill.

Now, Seamus was an old-country nationalist, a purist of the filthiest kind. To Seamus, Irish was Irish was Irish, or Irish was nothing at all. First to arrive, he stood guard at the door. Inspected every dish. Said this was an Irish stew night – only mutton, potatoes, and onions allowed. No Welsh Glamorgan sausage. And certainly no steak and kidney pie.

While his intentions were sound, so were forty-two screaming kids. Step up, Ida Bubar. She ordered all pots cleared away. She laid out a tablecloth – blindingly green, though delightfully patterned in shamrocks. "Now return every dish to the table," she said. "And bring the ones Seamus has stacked in the snow."

She sprinkled the dishes with marigold. Chawing a plug of some curious blend, she jabbered in abracadabras, "I see seven sprites on the table …"

"I see fourteen," cried young Frank McCready.

Ida packed a chew of a stronger blend. "I'm countin' nine, Frank. Catchin' up!"

"But I'm seein' eighteen now."

Ida sucked every crumb from her pouch. Minced them to mash, slurped down the syrup. Muttered long enchantments not heard since Merlin. "I can see twelve – but can't get no higher. So, hear me, you meat, fish, or fowl – whatever you are in these pots tonight, you will now have the flavour of Irish stew." She turned toward Seamus. "How say you, Seamus O'Neill?"

Ol' Cyclone looked up from his kidney pie. Squeegeed the drool from his beard. "'Tis the best Irish one ever had." Ida collapsed into her chair.

Supper over, the families bundled up, stepped out into... June – sunshine, birdsong and blossoms; horses nibbling clover in meadows of green.

Ida should've stopped with that first batch of sprites. She forgot about that Clydesdale's kick to Frank's head – his permanent case of double vision.

Observable. Repeatable. Verifiable.

But there's a fourth pillar in the scientific method: the pillar of *falsifiability*.

Huh? Just this: The theory you're proposing must be able to be disproved.

Here's how it works.

You have a theory: "2 hydrogen atoms and 1 oxygen atom will give me water." So, you mix them up, and, lo and behold, you get water. You try it again – the same result. Once more. Same thing. Every time. It's looking like your theory may ... well ... hold water.

So then you say, "I've got another theory: 2 hydrogen atoms and 1 sulphur atom will give me water, too." You mix them.

You get something entirely different. Your chemist friends tell you your theory stinks (hydrogen sulphide), but since you hold your theory to be true, you call this second result, water, as well.

You come up with another theory: "2 hydrogen atoms and 1 selenium atom will give me water." You mix them up; you get something different again. (At least your chemist friends have quit rotten-egging you on, lying there dead on the floor. Yeah. Hydrogen selenide—a gas that will kill you.) But since you're absolutely certain your theory is true, again, you call what you got, water.

See what's happening? It doesn't matter what goes into the mix, or what you get for an outcome, your theories have already declared that you shall always end up with water.

All of your theories are proven correct. They must be correct because you ended up with water every time.

It's the same with the theory of evolution. It can never be disproven. As follows:

The recipe calls for uniformitarianism — processes working consistently, and at the same rate over time. But, if the theory needs systems to speed up for a while, or to slow down for a billion years, or for mechanisms to disappear, only to re-emerge when called for — it's all good. We can throw whatever into the mix, and still call it 'evolution'.

The theory demands an increase in genetic information. But if science proves information is being lost, it doesn't matter. Stir that in, too. It's still 'evolution'.

The recipe demands transitional life forms to get from one kind to another kind. But if the fossil record shows the kinds haven't crossed over — that's no big deal either. Fine. We'll take it. It'll settle out somewhere. And we've still got our recipe called 'evolution'.

See? Throw in all the pluses and minuses you want. Toss in all the contradictions; everything becomes an ingredient. No matter what happens to the texture, to the colour, no matter how it changes the taste, you still get to call it Irish Stew.

A clear case of a theory held higher than science, evolution can never be falsified. Any variable can shift at any time. There's no possible way to get a negative result, one that would demand you toss your theory.

You just keep upsizing your pot.

Reflection 29 – Design

dayspring on summer meadow

… first blush
blends with the blackness
of space,
morning-swell's creep
imperceptible

til burgeoning hues
distending,
burst,
condense to that increscent
arc –
to that fingernail moon
of a sun
set to rise …

mackerel sky, salmon
streams pink to rose,
floods Dawn in her sanguine glow –
the yawn of Creation in three-quarter time:

the droop of pulled taffy, molasses
in mist,
fog-fingers steal
off to fill empty pockets;

leaves

sopped in water-weight, bow
in obeisance;
forced worship –
faith wrung from the science
of frontal condensation;

lone cricket's monophony
lost to the hollows,
his chirp, Love's Lament
on a ledger of flats –
no rests,
single notes held to a value of half;

cobwebs in crosswinds,
doilies-in-drip,
strung where the breeze funnels
bugs –
the death-rattle buzz of the fly,
the fight of the bumblebee,
caterpillars
en route to Butterfly Circle,
tied up in traffic –
rough ride for a highway of silk,
though the road
is
illumined now

— swag lines of ten thousand street-lamps —
dewdrops engorged with the sun.

We blind ourselves.

Go stand in the middle of your living room. Close your eyes. Spin around three times. Stretch out an arm, point a finger.

Okay? Now, open.

No matter the object you're pointing to, it was designed and built. No exceptions.

* Rocking chair – designed and built
* Coffee table – designed and built
* Desk, lamp, a remote control — each one, designed and built. Made the journey from a head to a skilful pair of hands. Head to handiwork, thought to wrought — any object you care to point to.

Now, go back to where you were spinning. This time, point through the window. Point to a tree, to an ocean. Point to a turtle, a bird.

Well?

Strange, eh?

It's funny, isn't it, how we so glibly divert our thought-stream; how we let our rationale take us only so far?

Supressing what so readily flows from thought. Refusing to acknowledge what's so clearly seen.

Maybe 'funny' wasn't the best word to have used there.

Reflection 30 - Authority

Sean was the boy they built woodsheds for. When it came to submitting to authority, he'd rather bend over than straighten up. He took his beatings with a smirk; left him with a bent set in concrete.

The village consensus was that Sean was stupid. Now, hold on a minute. Folks weren't trying to be mean — judging some supposed intellectual liability, trying to grind the guy into the ground. Wasn't that at all. They knew Sean was a whiz when he wanted to be. Hadn't he memorized The Highwayman in under four hours? The Wreck of the Hesperus in two? And he was the only kid around who knew the 3Rs: Relief, Recovery, Reform — the planks in FDR's New Deal.

No, Sean's brand of 'stupid' was the true kind of 'stupid' — as when one has the chance to walk a new path, but keeps to the road always taken.

Seems his life's motto was, "You're not the boss of me" — a mantra oft repeated when his dad dragged him off to the shed. "I'll do what I want." Or, "You can't tell me what to do."

He headed to Halifax the day he turned seventeen; grew tired of repeating Grade 8. The last contact his mother had was that one letter he'd mailed from Rockhead Prison. He'd needed ten bucks. She could spare only two.

She kept the letter hidden in a false compartment in the back of her bureau: Sean's letter, Wharton's, The Age of Innocence, and the slingshot she'd taken from Sean two days after his tenth birthday.

He wouldn't listen when she told him to stop shooting birds.

Every one of us holds to some authority.

Beneath every premise, every supposition, beneath every truth we hold or reject, lies the ground upon which our presumptions are built. The bases for what we hold as fact all rest upon something, or someone.

My parents told me that ...

I read somewhere that ...

But our teacher taught us ...

According to Time magazine ...

I heard the TV reporter say ...

We appeal to authorities to support statements we've made, beliefs we hold, or for justification for the things we do. We stand with, upon, or under, some authority somewhere.

"Donny, dear. Finish your peas."

"Yes, Mommy."

By Donny's action of eating his peas, he's choosing to submit to his mother's authority.

"Donald. Go pick up your bicycle."

"Oh, alright."

Donald stomps out to the driveway—under protest, to be sure—still, he makes the choice to put himself under the authority of another.

"Don. Give me my keys."

"Forget it. I'm taking the car."

Oh-oh. What just happened? For something *did* happen. (And it goes waaaaay deeper than teenage rebellion.)

Don's non-compliance (they won't let me call it 'disobedience' anymore) is more than a mere refusal to obey. And it's more than a choice made, too. While it *is* a refusal, and it *is* a choice, his action effects a shift in positions.

Don's rebellion pulls him out from under another's authority, and plunks him down on the throne. But his mom moves, too. Rather, she's moved by another—displaced. She's now underfoot—Don's insurrection, a usurpation.

And the ripples continue. Don's new position carries with it a new charter:

It matters no longer what anyone says, or what others expect of me, I'm my own person now. I determine what's right, what's wrong. I say what's good, what's bad — light, dark; sweet, sour; it all comes down to my say-so. I'm the judge of these things now — the arbiter of truth as I see it.

Well! Congratulations, Don, on your liberation: out from under the thumb. You can now keep all your resolutions, meet your every expectation, live up to all the standards you've set for yourself. Must be a wonderful thing, self-mastery.

What's that you say, Don?

This all has a hollow ring?

You keep falling short, even when the authority you're under is you?

Reflection 31 - Conclusion 1

There you have it, dear reader—an examination of real discrepancies between the observable evidence and the stated claims of several scientific disciplines.

Are they valid concerns? The answer lies in the journey. Any stumbles along the way? Any tiptoeing here, side-stepping there? Any sections where you stomped straight through?

Now, remember, I'm hiking here, too—on the path with you. And it has been a trek. Am I tired? Yes. Injured? You mean more than twisted ankles, and these skinned and bloodied shins? (Could someone hand me the ointment and gauze?)

Yes, I'm bruised, and not a little sad, for science is a wonderful, even intimate, thing. Left wounded, but wondering, are there not some weighty matters we need to consider?

And, no, I do not have all the answers.

But I sure have a heap of questions:

*If all life evolved from a chance formation of a single cell, how could that initial cytoplasmic division (a process requiring volumes of genetic information) have occurred *before* that genetic information had evolved, enabling such division in the first place?

*An intricate array of replication proteins is absolutely essential in the reproduction of DNA—DNA can't be formed without them. But the proteins themselves are coded in the DNA. How could that first reproduction have occurred, with both the DNA and the proteins being required concurrently?

*If the information stored in the DNA must be decoded by the ribosome in order for proteins to be built, how is it that the proteins required for building the ribosome are themselves encoded in the DNA? The info needs a decoder, yet the decoder's blueprints are contained in the info. Go figure.

*If each cell contains more than a thousand 500-page volumes of information, how could such life-enabling knowledge have risen spontaneously from life-lacking chemicals?

*How can a one-and-the-same gene code for two completely different functioning proteins? *Which protein do I make now?* There's more than chemistry here. Directive forethought is required: impossible through evolution, which, by its very definition, is a mindless, purposeless process.

*If all life evolved from a single, simple cell, why does the fossil evidence show complex forms in the earliest rock—their structures and organs complete, and (at one time) fully functional—yet show no transitional forms?

*If the theory of evolution demands a march ever onward and upward, why do the fossils show stasis—today's life forms clearly seen in the record of the earliest rocks? And why do the fossils show a greater number of phyla and species existing back then than exists today?

*With science showing natural systems having a tendency towards disorder, how can the evolutionary process claim to be going in the opposite direction?

*With the theory of evolution demanding an increase in genetic information, why does the science show that mutations result in a loss of the same; every child entering the world with one hundred additional mutations than either parent was born with; humankind hit with 500,000,000,000 new downward-driving changes in each generation?

*If mankind's evolution came about as a result of a 'hopeful monster progression' (a sudden, cross-kind appearance of a new life form), wouldn't two morphologically-diverse-but-compatible individuals (male/female) need to pop into existence in the same generation to carry on the line?

*Since animal fossils are formed from creatures that once walked upon, flew over, or swam about on this planet, why are there no living transitional-form animals walking, flying, swimming—themselves candidates for fossilization?

*If fossils formed from the slow accumulation of layering sediment, how is it that the evidence shows proof of rapid, destructive, cataclysmic burials?

*If specialized organs and structures develop through a series of random mutations, why would an incomplete, in-the-way, drag-me-down 'protuberance' of any kind, be seen to be advantageous; be the line chosen through which a natural selection of the ... fittest ... emerges?

*How can the claim be made that processes have always acted as they do now, when evolution demands abiogenesis? Evolution maintains that *protoplasm arose from non-living matter way back when, but that same process cannot be observed today.* So, is uniformitarianism in or out?

*How can there be a uniformity of natural processes, when 95% of fossils are marine invertebrates found in continental rock? Clam shells at the top of Mt. Everest? Whale fossils in the Andes? Only catastrophic conditions could have stockpiled dinosaurs. Where are those same conditions today? Something has changed (and, thankfully so), but this is not uniformitarianism.

*If carbon dating is only reliable to 100,000 years, how can it be used to prove ages of hundreds of millions of years?

*If various types of radiometric dating methods render a same-sample age difference of 3,000,000,000 years, why would I trust the results?

*If dinosaur bones are 65,000,000 years old, how can they contain material not fossilized: soft tissue, red blood cells, and DNA — this with a half-life of 520 years?

    These are reasonable questions in want of logical answers. No need to get one's ire up; we've posed no enraging queries here.

    But we do have eight chapters to go.

Reflection 32 – Conclusion 2

While the following statements are abhorrent, they are intrinsic in the theory of evolution.

Deny one point and the theory is lost. You cannot claim to accept what evolutionary processes have done, and at the same time deny what those processes do.

For as fills the wellhead, so flows the stream.

NOTE: I am NOT espousing these conclusions.

Nevertheless, they issue from logical sequence, and cannot be denied if the theory of evolution is true.

Follow the progression. Stop me if I veer from logic.

*First-life arose when a chance spark was kindled from a random mix of non-living chemicals.

*The subsequent progression from simple to complex life-forms involved time, chance, and a bit of luck — individuals possessing the most favourable traits emerging as progenitors of the species.

*It was from this first-life single cell passing through innumerable developments over vast ages, that humankind eventually stepped forth.

Good so far?

We continue.

*With all species arising from one common stock, then humankind too, is of animal source, endowed with animal nature.

*Mankind must therefore possess animal drives and appetites. Though each species may have a composite palette of urges unique to themselves, these drives are embedded in that species' nature.

*Therefore, it must follow that humankind—both through source and development—has the innate and existential right to live out its animal nature. And not only the right, but to live as such, its only expectation. To live as an animal is all an animal can do.

What defective system of reasoning would hold dictums demanding an animal live contrary to its nature, and then punish the poor creature for doing what its nature compels?

Dogs fight over food. We don't put them in jail.

Deer fight to the death for the choice of a mate, and the killer-victor walks away free. And not only free, but is rewarded with the spoils for behaving as his nature directed. The working out of his natural drives sees the deer sated, well-served.

Other animal species stab (hummingbirds), cannibalize (spiders), kill their own (wolves, monkeys); sharks eat sharks while littermates in the womb—all behaviours born of instinct. Shall we demand their transformation?

Leave man alone. Let him live as the animal he is, as the root of his nature dictates.

Now, while some may find these statements repugnant, they cannot be dismissed from want of logic. Objections will not stand.

I don't like the sound of all this. Man is just ... different.

If man is different, then he could not have come from animal stock. Objection lost.

But we're talking humankind here. Be reasonable.

You appeal to 'reason'? Is it reasonable to expect an animal's nature (interior) to obey a code of laws (exterior), laws which are contrary to his very being? Objection lost.

But, murder? Incest? Cannibalism? These are terrible things.

Who says they're terrible? And what are you doing, making

a judgment on another animal's nature? Does one animal tell another its natural behaviour is wrong? Sorry. Objection lost.

But we expect more from man.

But, why? Why expect more from one who's just another animal? These are all things animals do. Have you not already affirmed that man is of animal line? Again, objection lost.

Then, yes, okay; he's an animal. But what about community? Social order? What about acceptable behaviour? We are social animals — members of a society.

But are there no other social animals, others that live in community? Bees, ants, dolphins, whales. Are we to tell a society of wolves some things they do are wrong? Tell them to squelch their desires?

And, further: You are aware that if animals are confined, or their behaviours suppressed, they suffer all manner of woe? Emotional angst. Self-mutilation. No self-discovery. No personal fulfillment. To stifle self-expression does the individual harm. To restrict what's ingrained must surely be labelled the ultimate in animal cruelty.

Man is either fully animal (as the theory of evolution declares and demands), living by, in, and under animal ways and appetites, or, man is different from an animal, with higher things expected of him (which the solely physical succession of the theory can't allow).

Objections lost.

But the evolution stream flows on.

And while the bent of a river is to run,
silted, it fouls the sea.

Reflection 33 – Conclusion 3

From a single cell to the top of the food chain, how far humankind has come. Of all species which inhabit the planet, man has climbed the highest, has floated the furthest downstream.

He's done well for himself: for every other living thing, maybe not so much—both plant and animal kingdoms decimated in the ascent of man. What a dark demonstration of folly: attacking the Earth that conceived him, nurtured him; that grew him, and gave him birth. His womb, he's defiled; his nursery, he's poisoned. Earth's circuits and cycles, mankind has subverted—life, now on the edge of extinction, set to topple in global collapse.

No surprise. For this is man's impetus: *Survival of self at the ruin of all else* – this inbred aberration crowning him the species of renown; vainglorious, this attribute from the primordial swamp, this feature from the black lagoon.

But the grisly crawl of logic continues:

*If thousands of animal species have already disappeared, then animal-man's extinction is no big deal. What's one more?

*If the species on the verge of extinction is the one responsible for destroying the Earth, why would anyone want that species to continue? Shall we unite in our efforts to keep the Killer alive! This is idiocy of the nth degree. Let the Spoiler go.

*Think: What other animal species could have done a worse job than mankind has done?

That's worth repeating.

Here, now, standing at the brink of global ruin, what other animal species could have done a worse job overseeing this planet's welfare than mankind has done?

No other species could have done any worse.

So why all this fretting over mankind? Consider the evidence. Examine his track record. He's proven he's no good at running the Earth. I say, good riddance. And remember, he's only here by chance as it is; evolution avows no mind. Give the earth to the snakes, to the whales; to those radioactive-chewing bacteria that live in the dark; to those recently discovered plastic-eating microbes — they'd never run out of food. Let them take over in a terminal symbiosis: fattening themselves while cleaning up Mother Earth.

But, in the end, it really doesn't matter which species rules; that's the whole premise of evolution. Time will pass. Survival of the fittest will do its thing. A superior, more adaptable species will eventually rise to the top — just as man did once long ago. Who's next in line? Let them step up.

All this worry over mankind's extinction? I say, quit struggling. Hand over the reins.

Now, of course, all of this is terrible stuff (though not entirely facetious).But you can't fault the logic, not if man's just one more animal.

But hang on. The flow continues:

*How's the deer population this year? There's talk of culling the herd. Why would Natural Resources contemplate such a move? Simple. If you get rid of the sick and the weak ones, the overall health of the species improves. The same's true for all animal groups.

"All animal groups". So why are we spending vanishing resources propping up the sick and the maimed? Food, shelter, medicine — all wasted on those who are only weakening the species. Haven't you heard? Evolution declares it's the *survival of the fittest* out there. All species fare better with no draggers-down hobbling about.

And yet the species 'man' spends over half of all his resources on mending the weak ones? What a stupid animal man is. No wonder the earth is dying. He'd be far better off —

his chances of survival increasing a thousand-fold — if he'd forget the sick: direct all energies to buttress the strong.

I repeat, this is terrible stuff. But the reasoning and the logic are sound; both of which must, and do, flow from the theory of evolution. Man is an animal. Period. Let him live as the animal he is.

If these reasonings appear distasteful, these rationales, ridiculous, then here's a final consideration.

Remember evolution's hypothesis? It says that certain individuals in a species will evolve faster and further than others — progenitors from which new lines will emerge; those who will lead the species onward and upward to the next stage in their development.

Well, here's the full title of that 1859 book:

On the Origin of Species by Means of Natural Selection or **the Preservation of Favoured Races** *in the Struggle for Life.*

H'm. I wonder who the favoured races are?

Reflection 34 – So ... where to from here?

Michael Young was lost for fifteen days. His older brother, Edward, blamed their Aunt Ellen in Portland, Maine, three hundred miles from the Durham woods. She should not have sent that Thanksgiving gift, though he did render plaudits for her knowing the Canadian difference.

Edward had been raised for the wilds, taught how to live off the land – with the land – from the land. His dad had shown him how to read the wind; how to plot direction from the thicker side of trees, or from the moisture in a circle of moss; which plants were safe for the belly, which ones would poison the blood. He could read the pole star – could line the tangent from the tips of a crescent moon. Edward knew the way of the woods.

Michael, too, loved the outdoors, though he was the gentler soul. Not in the felling of trees, and not in the taking of game, his joys lay hidden in birdsong, in the burble of brooks, in the palette reflected in mill ponds. Leaf texture. Wildflower carpets. In the grain lines running through rock.

But then, October, 1914: that package from Maine – Aunt Ellen's gifts, with a note:

Happy Thanksgiving, dear nephews,

Edward, the money's for you. I know you're saving for a rifle. And Michael, I thought of you when I saw this compass at Rines' Brothers (just one of our large department stores here). May it keep you safe in your outings.

Love,

Aunt Ellen

The morning after the parcel arrived, Michael grabbed his notebook and compass, snuck out the back door and into the woods. Two hours. Twelve hours. He wasn't seen again for fifteen days.

*   *   *

He remembered his dad saying to bed down on high ground – away from the danger of swelling streams – and to hang your food high and away from your tent. Tent, bed, food: he would have died for any one of them now, and nearly did from the lack of the three. He wished he'd paid closer attention – maybe taken some notes, as he would do every day in his wanderings.

"Thanks, Aunt Ellen, for this faulty compass. Gives one reading from the left hand, another from the right."

*   *   *

Two weeks later, Michael staggered out of the woods – Michael, minus fifteen pounds. It would've been a heavier loss if not for the scattered offerings of withered blackberries – apples and cranberries, more sympathetic.

After a week of prescribed feedings and sleeps, a few friends gathered to hear Michael's story, to listen to him read from his journal. Each entry was dated – told what he'd seen for wildlife, and what he'd collected for rocks.

He never got to finish the first page.

"Dusk: Day one. Hungry. Made a bed of fir boughs – guessing some were spruce from this gum on my shirt. I found a strange stone, lustred, with sparkles ... ." He looked up from his reading. "Here, I can show you." He opened his bedtable drawer. "I found this near the – "

"Michael." Edward grabbed the rock from his brother's hand. "This is lodestone! It's magnetite, Michael – magnetic interference. You can't get a true bearing with this in your pocket."

My emphasis has been three-fold:

i- True science is empirical. It's based upon knowledge gained through the scientific method: observable, repeatable, testable, facts.

ii- Historical science is conjectural: interpretations of facts not based on direct observations, but on inferences drawn from extrapolations and assumed conclusions.

iii- My questions have come from within the realm of science itself. I've fired no rounds from mercenary guns—the battle joined solely between *What Science Can Prove,* and *What Science Assumes When It Can't.*

And the outcome of this inquiry? Picture an embedded journalist—left wounded on the field by the belligerents; tossed a few rations from a sack of conundrums (and those, of the assaulted variety).

So, where to now?

What other voices are out there? What other domains address the questions for the answers science yet seeks?

Perhaps the next (dare I say) logical step, would be to turn to the Scriptures. That the Bible has something to say about origins is a fact universally recognized.

At the very least, it's one more voice at the table.

The reader will remember two things:

i- What follows is not an exhaustive look into the source or the purpose of the Bible. We simply take the Bible on its claim: *All Scripture is given by inspiration of God, and is profitable for doctrine, for reproof, for correction, for instruction in righteousness.*

ii- We've seen that observational science and historical science are not the same thing: a great gulf divides the two. This difference must be maintained in the mind of the reader.

For to overlay the Scriptures with the speculations of historical science, leads only to a blurred confusion.

But if our touchstone be that of empirical science, a different picture comes into view.

Reflection 35 – A most unsavoury blend

The Nashwaak Watermelon Society no longer exists. Once boasting sixty-five members, the club has gone to the pits.

And how did such a novel lot go out of print? Charter member, Sadie Bell, wrote the following conflation in 1963, on what should have been the Society's tenth anniversary.

Here's Sadie's account, unedited, left swaddled in the wraps of her own running commentary. All parentheses are hers.

September 8, 1953, and dates following:

Jimmy Segee climbed the creek bank to the road that ran by the Taymouth General Store.

"Mr. Munroe!" Jimmy dropped his gill-stick of thirty-two trout, ran up the steps, and straight to the counter. "Mr. Munroe! Are those watermelons in your window? Real watermelons? Where'd they come from?"

"Yessir, Jimmy, they're real 'right enough, though I can't say who grew them. But with all this hot weather, and school on again, I figured you kids could use some cheering up."

"Watermelons! Wow! How much?"

"How many do you want? I can ... uh, maybe move a little on the price."

"Just one, thanks. How much?"

Now, Bill Munroe was a kind old soul, but with Jimmy's dad, Peter, being the local gunsmith, Munroe thought he'd shoot high. "How's forty-five cents?"

Jimmy crumpled into the magazine rack. Shinnied to a stand, then turned out a pocket. "I can give you ... this nickle?"

"Now, Jimmy-boy, I'm a fair man, but I've got to have – "

"And trout," cried Jimmy. "This nickle, and your pick of ten trout. Fresh from the hole." He pointed to the path that ended in the creek. "Take the ten best. Will that square us up?"

Munroe reeled in a breath, picked up his receipt book. He touched the pencil tip to his tongue, then scratched out some figures in mumbled hieroglyphics. "Okay. Deal. Grab your melon."

"Thanks, but could you pick one for me – the one that'll be sweetest come Saturday?"

Munroe laughed as he walked to the window. "Don't know if I can guarantee that." He reached to the front of the pile. "But ... maybe this one? I had it where everyone could see it."

Jimmy wrapped the melon in his jacket. "Wait til the guys see this. Goodbye."

"And goodbye to you, too, Jimmy ... as soon as I get my fish."

(So, the beginnings of our Watermelon Society: the flurry, the zeal, the passion, and maybe the hint of a hidden motive – Jimmy having turned out only one pocket.)

* * *

On his way home, Jimmy ran into Primella Moore. "Good day to you, James Segee."

"Hi."

(Primella had the sweets for Jimmy – had since that day he'd carried her to school. She'd collapsed from a 'sudden and completely immobilizing fainting spell' – that's how she described it. I say it's funny how it hit her just as Jimmy was walking by.)

"What's that hanging on your shoulder, James Segee?"

"Fish."

"The other shoulder," purred Primella, "though they both look so equally strong."

Jimmy laid down his fish stick and unrolled his jacket. "It's only the world's sweetest watermelon. Ever seen a real one?"

"Why … I don't believe I have, James Segee. What on earth shall we do with it?"

"I thought I'd …. We?"

"Certainly, 'we'. There's a slice there for every child in the valley."

"You're right, Primella. So at school tomorrow, tell everyone to be at my house on Saturday. 10 o'clock. And tell them to bring a nickle."

(That's another thing I'll say about Jimmy: always quick on his feet when stood against a wall.)

* * *

Six kids showed up: Jimmy, Primella (You're arriving an hour early? Have you no shame, Prime-Ella?), Dickie and Bobby McGivney, me (Sadie), and my cousin, Tommy Urquhart.

At the appointed hour, Bobby banged out two chords on the piano as Jimmy pulled back the parlour curtain. There sat the melon, perched on the hutch. Jimmy called the meeting to order.

"First, your dues." He held a Wedgewood platter in the face of each attendant. "A nickle, and more if you've got it."

"Do we get a receipt?" asked Dickie.

"Let's eat," said Jimmy. "Slices all around." He passed me the knife. "Sadie, will you do the honours?" He swept the coins from the platter, restocked it with melon and retraced his steps. "Save your seeds. We'll need them once they've dried."

Jimmy crunched down the last of his rind, then stood beside the piano. "And now, the reason we're meeting. I thought of a way we can make some money by – "

"By demanding unspecified dues?" asked Dickie.

"No. By designated collections. And today's money goes to paying me back for what you're eating right now. Is that okay with you … Dick?"

"I suppose." (Dickie was going to need watching. I told Jimmy as much.)

"Good." Jimmy continued. "I thought we'd start a watermelon society; sounds fancier than 'club' — maybe more adults will contribute. We'll go to the library — learn all we can about melons. We'll save some seeds for planting, keep some for creating knick-knacks. We'll have watermelon socials — charge double for watermelon punch. Sell tickets to plays we'll put on. Write stories and poems; sell our wares at the store, at church raffles — the Stanley Fair's coming right up. But we need to start off with a bang. I was thinking, Primella, everyone knows you're a painter — "

"You're so sweet, James Segee, although I wouldn't say ... everyone. Would you? Really?" (I wanted to step up and choke her. Had to settle for rolling my eyes.)

"Oh, yes, Primella," Jimmy gushed, "everyone knows how talented you are."

"And you are so kind, James Segee. So terribly kind." (Primella, can we move this along? Some of us have chores to get to. And thankfully the barn air hangs a lot fresher than what you've got us breathin' right now.)

"You could paint plantation scenes," said Jimmy. "Show the melons being harvested; folks would love it. We'll draw up some posters; tell everyone your paintings will be auctioned at the Fair, all proceeds going to the Nashwaak Watermelon Society."

Dickie raised his hand. "And what will the Society do with the money?"

"Buy more melons," said Jimmy. "If they're not in season, we'll buy pots, and soil; lumber and glass to build starter frames. Sell the plants. Get into full production; maybe start our own greenhouse. Lots we can do. Well?"

Jimmy made it sound like we'd all be millionaires. We drew up our Watermelon Pledge:

"I, ____, do solemnly swear that the watermelon is the sweetest fruit in the world. I will plant seeds, write stories, draw pictures, sell tickets, attend socials — whatever promotes the watermelon."

All six of us signed that day.

We had twenty-four members by the end of the year.

* * *

But there was a spy in our midst: seems our six-charter-membership held a fifth column. (No surprise to me – that prissy Primella, always leaned more toward plums. Her sketches looked nothing like melons.)

Jimmy gave her three written warnings. (That's five more than I would have given.) He told Primella if she didn't confine her drawings to what was prescribed in the pledge, he'd be forced to call a Watermelon Tribunal. (Do it, Jimmy. Do it. Spit her out of the club.)

But in spite of all the reprimands, Primella persisted in picturing plums.

A Council was convened.

* * *

Jimmy reminded everyone that the purpose of the Society was to celebrate the watermelon, and all members must be of a kindred spirit. (I would have worded it differently. 'Why let a person with a heart full of prunes remain in a Watermelon Club?' That's what I would've said.)

Primella was given the floor. (And her defense was like nothing I'd ever heard). "Honoured members of Council: What about the starving children in Africa – those with no access to watermelon? Have we become so enamored with this ... passion fruit, that we would deny the undernourished the right to die from a lack of other varieties? Shall they starve from a dearth of melons alone? And, I would ask you all to remember, my daddy's the fourth generation of being the only doctor in the valley. And many families here have outstanding accounts." *She reached into her pocket and pulled out a thick fold of paper.* "Shall I proceed to read the list?"

Jimmy jumped. "That shouldn't be necessary, Primella. We may have some options here." *He took a minute to do some ciphering.* "Okay. This should do it. Members of Council, to avoid the charge that the watermelon is the only fruit suitable for not feeding the masses, we have two choices:

i - Revise the Pledge: Open our membership to any type of fruit lover: apples, kumquats, Jordanian figs – full privileges to all appetitual preferences.

ii - Do nothing: Let Primella continue to paint plums.

After three minutes of informed discussion, a blended resolution was adopted:

"The Nashwaak Watermelon Society is now open to any and all fruit lovers who will solemnly swear never to speak of the primacy of watermelons."

So, now, thanks to the amendment, everyone in the valley could belong. Or not.

And that's why you'll find no Watermelon Society meeting on the Nashwaak today.

Sadie Bell – from sources

Many claim the answers relating to origins are to be found in a blend of the realms: *We'll overlay assumptive theories onto the simple declarations of Scripture. Take something from here, something from there, come up with a syncretism.*

While this approach might sound appealing (like you would do before swallowing a melon), it won't work.

This middle-of-the-road course runs smack into a wall of unresolvables.

Here's two of many:

i- The unresolvable problem of **Direction**

Evolution has mankind's journey beginning at the lowest of lows, and moving ever upward from there: from amino acids, to proteins, to cells; to sponge, to worm, to fish; to amphibian, to reptile, to bird; to mammal – to primate chimp; on to humankind: every step, higher and higher.

The Bible says man first appeared as a finished creation in the image of God. And from this lofty position, he fell.

Two different starting points, and with movements in opposite directions.

And I'm to hold both as true?

ii- The unresolvable problem of **Chronology**

There's a very clear sequence in Scripture:

Gen. 1:1-2, 5 - the earth was created on the 1st day
1:14-19 - the sun and moon were created on the 4th day
The Earth was here before the sun.

Gen. 1:11-13 – trees (in fruit, with seed) were created on the 3rd day
1:14-19 - the sun was created on the 4th day
Life was on the Earth before the sun appeared.

Gen. 1:16, 19 – He made the stars also: 4th day
The Earth was here before the stars were formed.

Gen. 1:20-23 – fish came forth on the 5th day
Fruit trees were here before fish.

While you're free to reject what these verses assert, and to dismiss literal 24-hour days (make them a billion years each, if you'd like), the chronology declared in the Scriptures is impossible according to evolutionary theory.

Discount the Biblical evidence if you choose, but don't claim these two opposing worldviews can be blended into one.

The statements of Scripture, and the assumptions of historical science:

Fusion? No.

Confusion? Plenty.

Reflection 36 – A fork in the road

Dr. David Moore could see the end coming – the last days of the horse-drawn carriage. Though buying a car would mean no more sleeps in the back of the buggy, he would not be labelled a Luddite.

So the good doctor took the plunge – twice. He was first in the valley to purchase a car, and second to install a phone.

But the car brought some real complications. His horse, Ether, a loyal companion of sixteen years, had to be put out to pasture – a painful separation, though both knew it wouldn't be long-term. With the doc pushing ninety, there'd be a glad reunion soon enough.

And Ether wasn't the only concern. Moore knew that with his failing eyesight, getting a handle on driving would put him on a steep learning curve – a precarious venture indeed, his house on a straight, level run from town.

His great-nephew, Garrett, insisted on riding shotgun. "Just for the first few days, Uncle Dave. A man your age will need help in making the jump from a horse to a car."

Garrett quit on their first run to town. Couldn't stop Doc yelling 'Whoa' whenever he stepped on the brakes.

But the biggest challenge concerned Dr. Moore's arrival times – most notably, on distant rural routes. While a car should have shortened his ETAs, he started showing up fifteen, twenty, thirty minutes late. And without a word of explanation.

Why was he now late for his appointments? He let it slip the day of the wedding – his great-granddaughter Annie's reception. Corner table. Awful music. Dark rumblings of war. There with his fishing buddy, Alowishus Brown, Doc poured out confessions and 'medicinal' brandies, swizzled with chasers of tears.

He told Alowishus that Ether – "Oh, my poor Ether" – was the more-seasoned traveller of the two. With his twenty-seven rural patients – many of them, miles apart – there was time for an overworked doctor to grab a nap in the back of the buggy.

The car ruined it all: Doc scratching his head at every fork in the road, his at-one-time navigator off munching hay.

Consider the disciplines that run throughout story.

Any story: Nick, a man (Anthropology) in a bloodied (Hematology) paisley and lace-collared shirt (Sociology), stands on a treed (Dendrology) basalt cliff (Geology). He screams as he jumps (Psychology). He hits (Gravity) the water (Oceanography). Inhales (Physiology) six litres of H2O (Chemistry). Alas, the poor man dies (Thanatology).

Many dimensions in many domains. And while a more focused examination might add some depth (sorry, Nick) to the story—i.e., the man's relationships with his co-workers could point to a motive for murder—each realm within the narrative has its limitations; it can only take you so far. The cliff's composition (Geology) bears no relationship to the volume of water in the lungs (Physiology). The height of the trees (Dendrology) can tell you nothing of the rate of acceleration in his fall (Physics). Water temperature (Oceanography) says nothing of why he jumped (Psychology).

Domains have lanes.

Once you step from the path of observable fact, you leave science back at the forks.

And this divergent path you're walking now? Call it what you will: Philosophy? Religion? Metaphysics? But with you having abandoned the seens and the testables—you now framing your views with assumptions—don't call this variant path of unprovables, science.

Reflection 37 – I'll have a large grapenut, please

We were on our way back from running the dogs; one of those July days that has every known species dragging its tongue.

"I miss the ice-cream trucks. You, Gramp?"

"Naw."

"Yes, you do. You like ice-cream."

"Only the good kind, Kid. None of this licorice ... tofu ... stuff."

She'd run out of cereal – asked me to drive her to the grocery store. I offered to share my personal stash, but ...

"Sugar Pops, Gramp? I eat only Kashi. Organic Cinnamon Harvest."

"Of course you do. Okay, then. Tomorrow – nine o'clock."

Hard time to be a kid; this Covid mess – dark, exhausting uncertainties.

"So, you're ... seven now?"

"I'm nine and a half, Gramp."

"Really?"

"Yup," she chirps, chin in the air. "Ten in January. If you're still alive, will you come to my party?"

"I'll try to hang on. And I'll come on two conditions."

"I know. Yes, you can bring the dogs; there's some awesome dog-friendly cakes now."

"And just how would you ...? Never mind. But I'll come ..."

"Yeah – if we're serving your flavour. C'mon, Gramp. Everyone likes my Heavenly Hash."

"I'm not an everyone, Kid."

"Oh, I know that."

We pull into the parking lot. "See? Crazy. People lined six feet apart, heads bowed, thumbs ... what? Scrolling? It's nuts."

She pulls up her mask. "Why don't you wear a mask, Gramp?"

"Kid," I point, "look at this face. If you had these features, would you keep 'em wrapped in gauze?"

"You mean like those dead guys in Egypt?"

"Smart girl, eh?"

"Yup. That's me." She loses the grin. "Have you ever worn a mask, Gramp?"

"Used to all the time."

"Really?"

"Sure. Every boy wanted to look like Jacques Plante."

"Who?"

"See what I mean? The whole world's collapsing. Time was when everyone knew Montreal's goalie."

"Another old hockey guy, Gramp? Someone born in the nineties?"

We go stand in line. She drops to a whisper. "Gramp?"

"Yeah?"

"Did kids have ... problems ... when you were young?"

She's a strong one, well-planted, but there's something on her shoulders. Slips through in her voice.

I take the lighter side — maybe stir some hope into the mix. "Oh, we had problems, alright."

"Like what?"

"Like whenever we found a nickle."

"Huh? Why was that a problem?"

"Seriously? Staring at that ice-cream menu board — choosing a flavour from those cardboard tubs? People pushing, telling me to hurry up — me trying to decide between grapenut and orange pineapple …"

"Eeeeeuu!"

"Yeah. It's never been easy … being a kid."

"Need a mask, sir?" the guard offers.

I feel Lia's eyes on me —

"Sir? Mask?"

— those doleful Betty Davis eyes.

"Sure," I sigh in surrender. "Thanks."

"Lean over, Gramp. Let me help." She beams at the guard. "It's his first time."

I squeeze the nose band. "How's it look?"

"Awesome."

We advance to the next green dot.

"Gramp?"

"Yeah?"

"Just so you know — this line we're in now?"

"Yeah?"

"I doubt there's orange pineapple at the end of it."

Ice-cream: more than a thousand flavours.

To say I don't care for sesame-broccoli-maplebud, is not to say I've ruled out every other kind. While I don't like 998 flavours, I do fancy orange pineapple and grapenut. Maybe coffee, in a pinch.

Think of it! In the ever-expanding universe of slurpy selections, three flavours I hold steady-state. And it's not just the clutching of the cone in the hand; ice-cream gives a grasp of deeper things.

Begin with the letter 'A' — that indefinite article of choice. 'A' is a preposition, a noun, an abbreviation, and 'A' is a prefix: means 'not', or 'without'.

Amorphous – no form, no shape

Abiotic – no life

Asymptomatic – no symptoms

Atheism – no God

Break this one down. The two parts are readily seen: prefix, 'a' - no; root, 'theos' – God. A-theism: no God.

The word gives its own definition.

An atheist IS NOT a person who DOES NOT BELIEVE in God; an atheist IS a person who BELIEVES there is NO GOD (a-theos).

That's a world of difference. It's more than a world of difference — for in the big, broad galaxy of beliefs, the atheist has chosen a flavour. He's made a selection from the menu board. And the belief he has chosen is ... to believe there is no God.

And that's fine. It's his choice — his right to do so. You'll get no argument from me. But at least be consistent with words and their meanings. Don't say an atheist does not believe, because he does. He believes there is no God.

He holds other beliefs, too. Now, I'm just guessing here, but companion beliefs might include: He believes there's no Heaven. He believes there's no Hell. He believes there are no angels. He believes there's no judgment. The list goes on.

And again, that's fine. Free country, as they say. But this list of things he believes — this ideology of his, however long or short ... well, do you want to hear something really scary?

There's another word for 'list of beliefs'; you can find it in any dictionary.

Religion: a set of beliefs concerning the cause, nature and purpose of the universe.

Religion. Now there's an ugly word if ever there was one.

Religion: 'A set of beliefs'.

You know — like what's on your list.

We all have a list.

Reflection 38 – Look at that!
I close my eyes and it disappears

When Michael Maguire carved this field from the forest, he was careful to spare this tree – this lone red oak, now with a nine-foot girth, and a broad, sweeping canopy. And though it shelters from the sun and the rain, it extends scant refuge from the wind.

And it does blow here.

Its standing speaks of foresight. The sapling looked nothing like the tree does now, one hundred and fifty years on. Maguire was thinking of distant days that day he granted a reprieve – moved to inaction by the prospect of blessings to come: to the cattle (they could chew things over while keeping cooler heads), and to couples (sipping iced tea in barn-red Adirondacks, angled just so for the touching of hands).

And who did sit under these branches – cooed, and cuddled, and spooned; caught a blossom bouquet on a drift from the orchard?

Someone was here. Some two were here – this heart, carved into the trunk. Whatever became of these lovers – SR and ... illegible, this other set calloused in scar? Sad, but love does that, too.

But this heart-line cut – once weeping, once tender with love's expectation – insensitive now, to touch: SR and ... someone. Did they go on in love? Marry? Build a home? Children? SR and ... maybe an 'I', or a 'J'? Could be a 'T'. SR and some one – some two, in love.

This field's lain seventy years fallow. I've seen one cut of hay in the last twenty years. Fallow, but not dead, this tree standing yet.

For an oak can live four hundred years –

even with cuts in a wounded heart.

Fearing to think keeps a person from thinking:

I think not,

therefore I can

hold what I've always held so.

Antithetical thoughts
might lead me
to truths I would rather not know.

Let's listen in on Michael's decision:

This sapling doesn't look like much. But if I spare this tree now, someday — years away — this oak will be a blessing to some farmer's cows.

He lets the tree live.

But landform is not all that changes. Michael's life takes on a flavour, is infused with an expectation. He sees the oak casting shadow on some tomorrow's herd. Someday. It will. He believes it, and lives in the prospect of such.

This is what it is to exercise faith — no need to dip the term in religion. Faith is the expectation that what we hold to be so, is, and will be; one day, aspirations will precipitate to substance.

Whatever we believe was (origin), is (purpose), and will be (destiny), our ideology frames and directs our outlook. Beliefs engender anticipation.

Michael's oak — that tree he's counting on to someday shade the cows; the one carrying his expectations? Too bad about wind, and root rot. Rough stuff — leaf blister, oak wilt, blotches of anthracnose; regiments of army worms, back-scratching bears; mice nibbling bark under snow. What's to say the oak tree will make it?

Faith is not always rewarded, for every belief held is not true.

It would be a terrible thing, one day to discover, beliefs we held as fixed had been rooted in shifting sand — beliefs in need of continual revision.

Maybe…scrutinize our lists?

Even an oak tree dies.

Reflection 39 – I didn't know that was in there

Since many hold science as the litmus of all truth, it's pertinent to ask: Does the Bible make any statements which stand with empirical science?

Consider the truths listed below. (These are verse references only. Readers can look further.)

Statements of factual science in the Bible:

* The Earth has a hydrological cycle

Ecclesiastes 1:7 - All the rivers run into the sea; yet the sea is not full; unto the place from whence the rivers come, thither they return again.

* The winds move in their courses

Ecclesiastes 1:6 - … it whirleth about continually, and the wind returneth again according to his circuits.

* The winds have a weight that holds the clouds with their raindrops in balance

Job 28:24-25 – he … seeth under the whole heaven; To make the weight for the winds …

37:16 – Dost thou know the balancings of the clouds, the wondrous works of him which is perfect in knowledge?

* The Earth is a globe

Isaiah 40:22 – It is he that sitteth upon the circle of the earth, and the inhabitants thereof are as grasshoppers;

* The Earth is suspended in space

Job 26:7 – He hangeth the earth upon nothing

* The stars are too numerous to count

Jeremiah 33:22 – *As the host of heaven cannot be numbered …*

* No two stars are alike

1 Corinthians 15 – *for one star differeth from another star in glory*

* Animals reproduce within their kind

Genesis 1:24 - *And God said, Let the earth bring forth the living creature after his kind, cattle, and creeping thing, and beast of the earth after his kind: and it was so.*

* Blood is vital to life

Leviticus 17:11 – *For the life of the flesh is in the blood*

* All people groups are part of the same biological 'race'

Acts 17:25-26 – *… seeing he giveth to all life, and breath, and all things; And hath made of one blood all nations of men for to dwell on all the face of the earth*

* The universe is moving towards thermodynamic equilibrium – the direction of, and, if not for the Lord's intervention, the necessary end of natural processes.

Hebrews 1: 10-12 – *And, Thou, Lord, in the beginning hast laid the foundation of the earth; and the heavens are the works of thine hands: They shall perish; but thou remainest; and they all shall wax old as doth a garment; And as a vesture shalt thou fold them up, and they shall be changed: but thou art the same, and thy years shall not fail.*

* Carriers of infectious diseases were quarantined during the communicable stage of the illness

Leviticus 13:13 - *Then the priest shall consider: and, behold, if the leprosy have covered all his flesh, he shall pronounce him clean that hath the plague: it is all turned white: he is clean.*

> :14 - But when raw flesh appeareth in him, he shall be unclean.
>
> :15 - And the priest shall see the raw flesh, and pronounce him to be unclean: for the raw flesh is unclean: it is a leprosy
>
> :46 - he shall dwell alone; without the camp shall his habitation be.

Though written 2000-3500 years ago, these are clear declarations of scientific facts. They've not been invalidated by 'recent discoveries'. They're in no need of the 'latest update'.

Granted, the Bible is not a textbook filled with scientific formulas, but the statements it makes concerning observational science hold true.

Moreover, many of these Biblical facts have been in agreement with science *before* science knew these facts to be so.

Sorry. If you want to read about the Earth resting on a stack of turtles, being created by a raven, or emerging from a cosmic egg, you'll have to look elsewhere. The Scriptures don't deal in fables.

But inventions are certainly out there.

And they're not far away.

We're told we have to overlay the Bible with assumptions to make it agree with science. Not so. Any science spoken of in the Bible, comes through just fine on its own.

We're told we can't hold both science and the Bible as true—the twain shall never meet. Not so. The two walk hand-in-hand.

While it's certainly true that theories based in historical science—based on assumptions and extrapolations—don't merge with the claims of Scripture, empirical science and the Bible have always been in agreement.

Is it such a strange thing to ask: If the Bible stands true in matters of science—the realm into which the eyes can see—does it not merit a listen when it speaks of the spheres the eyes can't perceive?

Epilogue

Thanks for walking with me through these Nashwaak reflections—born from the river, from this high field; this sunset home of the Hermit thrush, where ethereal echoes haunt silent wood; where my dogs run free while I stand fixed in thoughts of other worlds.

My purpose has been to recount that all is not as it seems.

Don't misunderstand me. This discourse is no denial of fact. The evidence is incontrovertible. A bone is a bone; a rock, a rock; and a fossil is certainly what was. The evidence *is* irrefutable.

But, the interpretation of that evidence, the assumptions upon which those interpretations are forged? These are different matters entirely. And this is where the problems lie: between what I can see, and what I'm told I must assume—this chasm into which true science is let fall.

It's from this breach that these questions have come.

Are we not to inquire? Are we not to ask *How can it be*?

To question is the very heart of science. It's in the light of true science, in the light of the evidence, I ask: *Why are things explained (or ignored) as they are?*

Do a little digging.

At least, kick a few stones.

For while there may be diamonds in them thar hills, you're sure to find a few at your feet.

Acknowledgements

Thanks to both poet and scientist:

One whose soul carries to mystical realms — away on the wings of a word;

One whose mind binds to empirical fact — tethers to the weight of the tangibles.

We need you both —

life but a venture,

a run on those rails

seen melding in distant diffractions of sand,

but never do meet —

ever leave us short of the ranges.

About the Author

Neil Sampson is a horticulturist who inhabits the worlds he hears in the whisperings of abandoned apple trees. Grafting poetry with prose, he fixes the science of plant physiology with the faith typified by the seed. An historian from way back who wishes he'd stayed there, you can find Neil on Twitter: @neilsam567

RESOURCES

DEVOTIONAL SPECIAL
GET ALL FOUR DEVOTIONALS AT ONE LOW PRICE !

Read one and give one away
4 FOR $35.00

Inspired Evidence

This extensively illustrated devotional provides a daily reminder that the truth of the Bible is all around us. Arranged in an enjoyable devotional format, this 432 page book starts each day with yet another reason to trust God's word. There is no conflict between science and the Bible – share that truth with others.

A Closer Look at the Evidence

This book offers unique evidence, primarily scientific for the existence of our Creator. It is organized into 26 diffe ent subject areas and draws from over 50 expert sources. Each page highlights awe-inspiring examples of God's incredible handiwork.

Have You Considered

This hardcover book is the third in a series of unique devotions presenting evidence for every day of the year that demonstrates it is beyond ANY reasonable doubt that we have a Creator.

This book is written in a style that anyone will understand and appreciate; extensively illustrated and meticulously documented.

Without Excuse

Our fourth creation devotional with scientific evidence supporting Biblical truth for every day of the year. A great resource for the entire family.

OTHER CREATION BOOKS
BY SEARCH FOR THE TRUTH MINISTRIES

Brilliant

An extensively illustrated book of the 50 best evidences for creation. Divided into biology, geology, and cosmology, each 2-page spread shows both the evidence for creation and the problematic evolutionary response. (8.5" x 11", 128 pages, hardcover)

Censored Science

How do all world cultures fit into a Biblical time frame of ~ 6000 years? An extensively illustrated look at the brilliance of mankind from the beginning of time. (8.5" x 11", 128 pages, hardcover)

A Christian's Guide to Refuting Evolution

This full color, extensively illustrated book is filled with activities, video links, and demonstrations. The book systematically reveals that the "best" evidence for evolution is riddled with contradictions and misconceptions. (8" x 11", 112 pages)

Science Shadowlands

40 short intriguing reflections on the evidence for creation that blends poetry, history, and everyday observations with science. A TOTALLY unique look at the evidence for creation. (5.5" x 8.5", 128 pages)

SEE ALL OF OUR RESOURCES AT WWW.SEARCHFORTHETRUTH.NET

CREATION CURRICULUM:

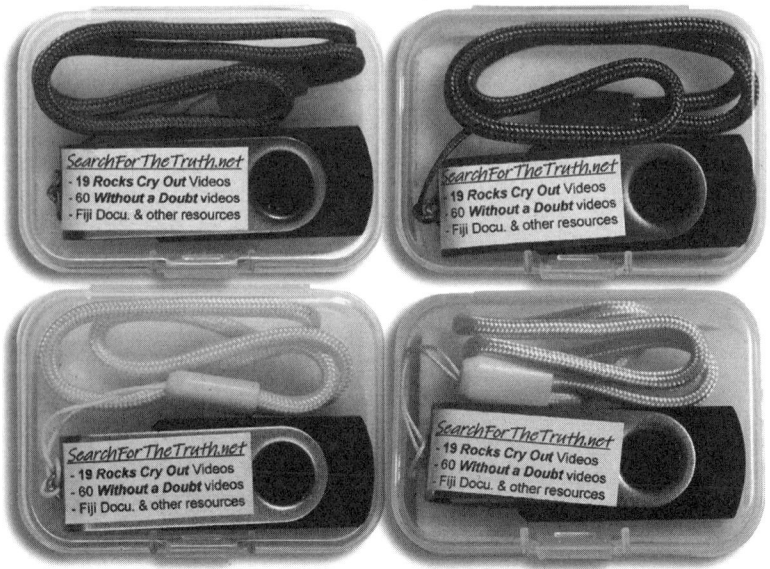

THE ROCKS CRY OUT CURRICULUM

Bring the most visual, interactive, and relevant series on the evidence to creation to your church, fellowship, or youth group! Filmed at locations across America with video illustrations and animations, these lessons are not a boring technical lecture.

These 45 minute classes enable the non-scientist to bring the evidence for biblical creation to their home or church. This curriculum uses short, personal narrative-style teachings to connect God's Word with science and history, i.e "the real world". Leaders guide included with each set.

Perfect for small group, home school, or Sunday school groups of all ages, The Rocks Cry Out show how EVERY area of science confirms iblical Truth.

See all of our resources at www.searchforthetruth.net

SEARCH FOR THE TRUTH
MAIL-IN ORDER FORM
See more at www.searchforthetruth.net

Call us, or send this completed order form (other side of page) with check or money order to:

Search for the Truth Ministries
3255 Monroe Rd.
Midland, MI 48642
989.837.5546 or truth@searchforthetruth.net

PRICES

	Item Price	2 - 9 Copies	10 Copies	Case Price
DEVOTIONAL SPECIAL (4 books)	$45.00	-	Mix or Match	-
Have You Considered (Hardback)	$13.95	$8.96/ea.	$8.00/ea.	call
Explore the World (Hardcover)	$13.95	$11.95/ea.	$8.00/ea.	call
A Closer Look at the Evidence (Hardback)	$13.95	$8.96/ea.	$8.00/ea.	call
Without Excuse (Hardback)	$13.95	$8.96/ea.	$8.00/ea.	call
Censored Science (Hardback)	$16.95	$11.95/ea.	$8.00/ea.	call
Brilliant (Hardback)	$16.95	$11.95/ea.	$8.00/ea.	call
Search for the Truth (book)	$11.95	$8.96/ea.	$6.00/ea.	call
Christian's Guide to Refuting Evolution (Softcover)	$12.95	$9.95/ea.	$7.00/ea.	call
Science Shadowlands (Softcover)	$9.95	$8.00/ea.	$6.00/ea.	call
Inspired Evidence (Softback)	$11.95	$8.96/ea.	$6.00/ea	call
Rocks cry out (Flash drive)	$45	-	-	call

MAIL-IN ORDER FORM

RESOURCE	Quantity	Cost each	Total
DEVOTIONAL SPECIAL (4 books)			
Explore the World (Hardback)			
A Closer Look at the Evidence (Hardback)			
Have You Considered (Hardback)			
Without Excuse (Hardback)			
Censored Science (Hardback)			
Brilliant (Hardback)			
Search for the Truth (book)			
Inspired Evidence (Softcover Book)			
Christian's Guide to Refuting Evolution (Softcover)			
Science Shadowlands (Softcover)			
Rocks Cry Out (Flash Drive)			
Tax deductible Donation to ministry			
		Subtotal	
Normal delivery time is 12 weeks	MI residents add 6% sales tax		
	Shipping add 20% of subtotal		
For express delivery increase shipping to 25%	TOTAL ENCLOSED		

SHIP TO:

Name: _____

Address: _____

City: _____

State: _____ Zip: _____

Phone: _____

E-mail: _____